Advances in UMTS Technology

INNOVATIVE TECHNOLOGY SERIES
INFORMATION SYSTEMS AND NETWORKS

Advances in UMTS Technology

edited by
JC Bic & E Bonek

HPS
HERMES PENTON SCIENCE

First published in 2001 by Hermes Science Publications, Paris
First published in 2002 by Hermes Penton Ltd
Derived from *Annales des Télécommunications*, Vol. 56, no. 5-6, GET,
Direction Scientifique, 46 rue Barrault, F 75634, Paris, Cedex 13, France.
www.annales-des-telecommunications.com

Hermes Penton Science
120 Pentonville Road
London N1 9JN

British Library Cataloguing in Publication Data

A CIP record for this book is available from the British Library.

ISBN 1 9039 9614 7

Typeset/Design by Jeff Carter
197 South Croxted Road, London SE21 8AY
Printed and bound in Great Britain by Biddles Ltd, Guildford and King's Lynn
www.biddles.co.uk

Contents

Foreword

In recent years enormous research effort has been devoted all over the world to specify, create and develop efficient radio interfaces and access network architectures in order to provide new services. Research laboratories, mobile operators, manufacturers, regulators have all contributed to the definition of a world-wide system. This so-called third generation mobile system is now coming to reality in Europe and Japan by the name UMTS (Universal Mobile Telecommunication System). The main features of UMTS are now well known:

- Spectrum efficient radio interfaces based on spread-spectrum and CDMA techniques, and sophisticated modulation and coding methods offering high capacity.
- Large bandwidth enabling broadband services with bit rates several times larger than enhanced second-generation systems, even if the 2 Mbit/sec bit rate per user would likely be limited to picocells.
- Ability to interconnect with IP-based networks, paving the way to truly fixed-mobile networks convergence.
- Flexibility of mixed services with variable data rates, providing a wide range of services from low-rate speech to interactive multi-media communications.

Now one of the most exciting challenge for the coming years is the deployment of these complex networks both from technical and financial viewpoints. Even if the planning is not so optimistic as it was one year ago, operations will certainly begin in 2002.

New services are crucial for the success of UMTS. Although the general service principles are stated (Open Service Architecture), the "killer application" is still well kept in the drawers of operators and manufacturers, and that is why this aspect is not deeply investigated in this publication.

In parallel with the implementation of the standards, research especially on the air interface is still proceeding at a rapid pace for even better capacity, quality and flexibility with enhanced transmitters/receivers.

This publication will address several issues related to UMTS emphasizing future evolution to improve the performance of Third-Generation Wireless Mobiles on the way to Fourth Generation. The contributions come from academic scientists, manufacturers and operators.

The first contribution, "Third generation mobile systems UMTS/IMT-2000" by J.-P. Charles describes the process that lead to UMTS in different standardization bodies, ITU, ETSI, 3GPP, and gives an overview of the resulting main characteristics for radio interfaces, network architecture and service principles.

The second chapter " Improvements in W-CDMA: Principles and Experimental Results" by M. Sawahashi, K. Higuchi, S. Tanaka and F. Adachi, reviews several critical aspects of the radio interface, channel code structure, spreading code assignment, rate matching and diversity. It proposes new techniques such as interference cancellation and adaptive antenna diversity for enhancing link capacity. Laboratory and field trial results illustrate the improvements provided by these techniques.

New access methods called MC-CDMA are introduced in the third chapter "Multicarrier CDMA Techniques for Future Wideband Wireless Networks" by M. Hélard, R. Le Gouable, J.-F. Hélard and J.-Y. Baudais. MC-CDMA combines code division techniques, DS-CDMA type, and multi-carrier techniques, OFDM type, methods. Their advantages in terms of capacity are demonstrated in the context of an UMTS environment. MC-CDMA turns out be a promising candidate for UMTS evolution.

The fourth chapter " Interpretations and Performances of Linear Reception in Downlink TD-CDMA and Multi-sensor Extensions" by L. Ros, G. Jourdain and M. Arndt focuses on modelling the multi-user TD-CDMA UMTS downlink channel and analyses the performance of multi-user detection in various indoor and vehicular environments, highlighting the benefits of joint detection and diversity reception.

Performance of smart antennas is investigated in the fifth chapter "Smart-antenna Space-time UMTS Uplink Processing for System Capacity Enhancement" by T. Neubauer and E. Bonek. Space-only and space-time processing techniques in the FDD mode with different service mix and system loading provide enhanced capacity by a factor of 2.5 or greater, depending on the mix of traffic services and system loading.

Deployment questions are addressed in Chapter six "Radio Network Planning Process and Methods for W-CDMA" by J. Laiho and A. Wacker. It stresses traffic profile and radio access technology as the most significant challenges for system dimensioning and radio network planning for a third generation W-CDMA system. Coverage is cell and service specific as opposed to second generation networks. Static radio network planning simulator results are compared to those of a dynamic simulator and are shown to be adequate for planning purposes.

The main characteristics of a versatile real-time test platform are described in the seventh chapter "An Open Software-radio Architecture Supporting Advanced

3G+ Systems" by C. Bonnet, G. Caire, A. Enout, P. Humblet, G. Montalbano, A. Nordio, D. Nussbaum, T. Höhne, R. Knopp and B. Rimoldi. Such test-beds are essential to try out and to validate new techniques proposed for the evolution of UMTS. The platform presently implements the physical layer of the UMTS/TDD mode, but could be extended to include new features such as multi-user detection or multiple antenna signal processing.

Finally Chapter eight is "Wireless Communications +++" by R. Steele, where the author expresses his views on the possible evolution of wireless networks. After recalling the recent past of second and third generation mobile systems, new concepts such as High Altitude Platforms, body-LANs, software agents, are discussed in the prospect of future wireless communications.

The editors would like to express their sincere thanks to all the contributors to this book.

J. C. BIC
Ecole Nationale Supérieure des Télécommunications,
France

E. BONEK
FTW, Forschungszentrum Telekommunikation Wien,
Austria

Institut für Nachrichtentechnik und Hochfrequenztechnik
Technische Universität Wien,
Austria

Chapter 1

Third generation mobile systems

UMTS/IMT-2000

J.-P. Charles

France Télécom R&D

I. Introduction

With third generation mobile systems, the world of mobiles will enter the era of multimedia. The stakes are considerable: around 2010, mobile traffic should be equal to that of fixed telephony. The convergence of mobile and Internet worlds, the strong dynamics of innovation, and the reduction of costs in these domains will open new opportunities for multimedia services. These systems could be brought into service as early as October 2001 in Japan, and around 2002 in Europe, in new frequency bands around 2 GHz.

I.1 Support of mobile multimedia services

The subscriber, at the beginning of the twenty first century, will use one or several mobile terminals (Figure 1) for different kinds of communications: the classical mobile phone, the pocket videophone, and the mobile PDA to manage diary, transportation, email, and to receive multiple information. With his portable PC, he will be connected to his company's intranet, and will benefit from videoconference service and all facilities needed to work outside his office.

***Figure 1**. Mobile multimedia terminals for UMTS.*

Several specific applications will use the capacities of UMTS systems to provide data, images or even videos: video medical diagnosis, reporting, proximity services, remote control, information, and driving guidance. Professionals' needs will also be satisfied through access to different means of telecommunication. UMTS will provide true mobile offices, even in vehicles.

Beyond professional use, the reduction of costs will lead to the generalization of these personal multimedia tools, the use of which should gradually extend to a large customer base following mobile telephony. Young people will spur the development of this market through their needs for games, education, sports. Thus, by the end of 2004, according to a number of studies, there will be 120 million multimedia mobiles out of a total of 1.1 billion subscribers in the world, and 4 out of 10 Internet users will also use mobile access to Internet at that time. Concerning data rates, UMTS is expected to offer up to 2 Mbit/sec, whereas GSM/GPRS can only support around 100 kbit/sec.

I.2. UMTS: a global mobile system

UMTS will offer a service of universal mobility, based on the success of GSM. It will be possible to access the same service independently of the environment: home, office, street, car, train. It will therefore be necessary to offer a great diversity of radio coverage schemes, from macrocells to picocells for indoor usage. The introduction of roaming agreements between UMTS operators will extend the geographic zone where the subscriber can access the mobile network. As UMTS will be largely adopted by existing GSM operators, but also by others which were not initially part of the GSM community (in Japan, for example), UMTS subscribers will be able to use their terminals in more countries.

Although the existence of other third generation systems will limit the ability to roam among the different systems, the fact that UMTS has been developed to ensure backward compatibility with GSM will be a key factor for the future, allowing a smooth transition between these two systems.

I.3. Migration from GSM to UMTS

The progressive migration of GSM networks towards UMTS appears to be essential to preserve the considerable investments already made in second generation mobile systems and to minimize the cost of introducing UMTS. To spread out the investments, UMTS will be deployed at the beginning in "islands" and GSM will ensure the continuity of service on the whole territory, but with limited services (voice and low data rates). This scenario is based on the existence of dual-mode GSM/UMTS terminals when the first UMTS networks are launched in Europe. For existing GSM operators, an upgrade of their core network will be possible since the UMTS core network is an evolution of GSM/GPRS, but they will have to deploy a completely new radio access network. Several thousand UMTS base stations will be needed to offer national coverage with high data rates and reuse of existing GSM radio sites will be a key issue to deploy rapidly.

II. International research and standardization context

II.1. Main players

In Europe, the development of a new mobile system has been largely based on research programmes launched by the European Commission in the early 90's. Japan followed another direction: most 3G developments were financed by mobile operator NTT DoCoMo. Japanese industry supported this R&D effort in order to develop a new standard and take the lead in this very competitive market. European manufacturers (Nokia, Ericsson) took part in this effort which led to the establishment of a common solution between them and Japan. A compromise was reached when ETSI (http://www.etsi.org) was looking for candidates for its third generation mobile system (UMTS). As in the United States, a large part of the frequency band allocated by WARC 92 (World Administrative Radio Conference 1992) for the IMT-2000 systems is currently used by second generation systems (PCS personal communication systems); it is thus not surprising to note that the American proposals for IMT-2000 often correspond to evolutions of existing second generation systems in order to maintain backward compatibility with them.

In this context, the standardization activities led within the regional (ETSI for Europe, TTC and ARIB for Japan, TIA and ANSI for the United States) and international organizations (ITU-R and ITU-T) developed with increasingly close contacts, as a certain convergence among the proposals took shape (in particular between Europe and Japan). In Europe, it is necessary to highlight the strong position of lobbies in third generation standardization (GSM Association, UMTS Forum) which are striving to federate, as far as possible, the stances of GSM operators and manufacturers. The regulation authorities play, in the same way, a fundamental role for the use of the spectrum identified by the WARC 92 and for the attribution of UMTS licences.

II.2. The standardization of third generation mobile systems

II.2.1. Standardization in ITU

The standardization of third generation mobile systems emerged in the ITU with the ambition of defining a global standard which would replace the existing systems. There could be no global mobile system without a common spectrum for all regions, therefore work on third generation systems really started once the WARC 92 had identified new frequency bands for IMT-2000 (Figure 2). This system, initially called FPLMTS (future public land mobile telecommunications system), then IMT-2000 (International mobile telecommunications) was expected to be launched at the beginning of 2000 using all or part of the spectrum identified around the 2 GHz band. This system was expected to offer high data rates, multimedia services, and global roaming.

Today, standardization harmonization on IMT-2000 is conducted in ITU-R/WP 8F (http://www.itu.int/imt) for the radio interface and in a new commission recently created in ITU-T for the signalling and networks aspects. The borders between the two entities still remain fuzzy for the protocols of the radio interface, taking into account the distribution of the activities between the two

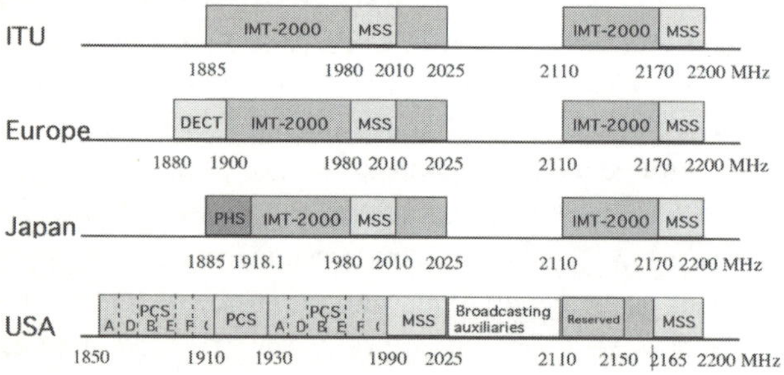

Figure 2*. IMT-2000 spectrum.*

sectors of ITU: standardization in ITU-T, Radiocommunications in ITU-R. It should be noted that the ITU development sector (ITU-D) is highly interested in IMT-2000 because many developing countries are waiting for such a technology to provide access to high data rate services with limited infrastructure.

In November 1996, the ITU-R approved the selection methods for the IMT-2000 radio interface. A call for candidates was then launched in March 1997, with June 1998 as the deadline to submit proposals for the IMT-2000 radio interface. Some technical evaluations were given at the end of September 1998. However, the ITU-R could not establish a consensus on any one of these proposals and, as a result, five different solutions were adopted in November 1999 :

- CDMA 2000 (evolution of the American CDMA IS-95 solution originally developed by Qualcomm);
- UMTS/W-CDMA (one of the UMTS modes supported by NTT DoCoMo, Nokia and Ericsson, and developed by the 3GPP);
- UMTS/TD-CDMA: second UMTS mode supported by Siemens. This mode is also developed by the 3GPP and it also includes a specific option developed for China;

- UWC-136 (evolution of the American solution ANSI-136 or D-AMPS); this solution integrates an evolution of GSM called EDGE (Enhanced Data rates for GSM Evolution);
- DECT developed by ETSI.

It is primarily the existence of different second generation systems (GSM, IS-95, D-AMPS) which prevented a greater convergence between these various solutions. The operators wished to preserve at least a part of the investments already made in the infrastructures while ensuring progressive migration towards the third generation.

II.2.2. Standardization in ETSI and 3GPP

In 1991, ETSI created technical sub-committee SMG5, to develop a third generation mobile system called UMTS (Universal Mobile Telecommunication System). This sub-committee was part of the technical committee SMG in charge of standardizing GSM in order to facilitate the migration of GSM towards UMTS. During the first years, this sub-committee co-ordinated the European positions for the ITU meetings. When ETSI had decided to propose a solution for IMT-2000, it became necessary to adopt a more flexible organisation to better define the European solution which would be proposed. Therefore, standardization activity on UMTS was distributed throughout the existing GSM technical sub-committees. The first stage of the standardization process for UMTS was to define technical requirements for the radio interface, mainly based on the work done in ITU-R, and the selection process. This process was launched at an ETSI conference in December 1996, during which various solutions were presented. Among these proposals, three solutions prepared by the European project acts/frames were presented (France Télécom R&D was part of this project).

France Télécom R&D was one of the rare participants to compare technically the various solutions in competition. After a vote, during an extraordinary meeting of the SMG technical committee in January 1998, a compromise was found based on two harmonized modes: W-CDMA [1] and TD-CDMA [1, 4]. W-CDMA was adopted for the FDD mode (Frequency Domain Duplex, i.e., one frequency per transmission direction) and TD-CDMA for the TDD mode (Time Domain Duplex, i.e., time-division multiplexing of the two directions on the same frequency). This mixed solution offers the advantage of allowing a complete use of the frequency bands allocated to IMT-2000: the FDD mode being used in priority in the paired bands and TDD mode in unpaired bands. This compromise was then submitted to ITU-R as the European proposal for the radio interface of IMT-2000.

The adoption by ETSI of FDD/W-CDMA opened the doors for an agreement with Japan, about to adopt this technology for its own third generation mobile system. Discussions among various standardization organizations: ETSI for Europe, TTC and ARIB for Japan, TTA for Korea, T1 for the United States, led to the creation, in December 1998, of a partnership among these organizations called 3GPP (third generation partnership project). This forum (http://www.3gpp.org) developed the technical specifications for UMTS. Then, these specifications were adopted as standards by the different national or regional standardization bodies.

III. Radio Interface

III.1. Objectives

Some of the objectives and constraints were defined before the design of the UMTS radio interface. These objectives strongly influenced the choice of the parameters of the various proposals, and it is necessary to point them out.

The UMTS radio interface was built to support a broad range of different services, with higher data rates than those offered by second generation systems (GSM, IS-95, PDC,...) (see Table I). UMTS offers circuit switched or packet switched mode services, with a maximum data rate depending on the environment and the speed of the mobile. Services with variable and asymmetrical data rates (between uplink and downlink) will be supported in an efficient way. Table I gives some performances: binary error rate (BER), delays for different types of services.

UMTS will be deployed in a multilayer cellular network, with macrocells (0.5 to 10 km) for overall coverage, microcells (50 to 500 m) for hot spots, and picocells (5 to 50 m) for indoor coverage. Handover will be ensured in a transparent way for the user, without any perceptible cut or degradation of quality.

UMTS will use spectral resources in an efficient way, by adapting the protection of the transmitted data to the radio channel. It will be necessary to optimize capacity and coverage. At the beginning, coverage will be the main goal of UMTS operators and then, gradually as the traffic increases, it will be necessary to increase capacity.

Planning of UMTS networks will be carried out if possible using automatic procedures. However, as for CDMA, coverage and capacity are closely linked, operators will need to use suitable radio planning tools in order to guarantee their customers the radio coverage, quality of service and data rate they expect.

The need for coexistence with second generation systems, and in particular with GSM in Europe, represents an additional constraint for UMTS. For that, it will be necessary to provide dual-mode GSM/UMTS terminals when UMTS networks are launched in Europe. Those terminals will be able to support handover between GSM and UMTS, which will allow progressive deployment of UMTS.

Table I. Performance requirements for UMTS.

Environment	Real time services		Non-real time services	
	Max bit rate	Delay/BER	Max bit rate	Delay/BER
Rural ($v \leq 500$ km/h)	144 kbit/sec	delay 20 – 300 ms BER 10^{-3} – 10^{-7}	144 kbit/sec	delay 150 ms in 95 % of the cases BER 10^{-5} – 10^{-8}
Urban ($v \leq 120$ km/h)	384 kbit/sec		384 kbit/sec	
Indoor and microcells ($v \leq 10$ km/h)	2 Mbit/sec		2 Mbit/sec	

III.2. The radio interface chosen by ETSI and developed by 3GPP

As indicated above, the solution adopted by ETSI in January 1998 is based on two harmonized modes: FDD/W-CDMA [1] for the paired bands and TDD/TD-CDMA [1, 4] for the unpaired bands.

In the compromise adopted by ETSI, it was also stated that the UMTS system could be deployed using only 2 × 5 MHz band, and that the selected parameters would ensure harmonization with GSM and dual-mode operation FDD/TDD while maintaining the objective of a low-cost terminal.

FDD mode is appropriate for all types of cells, including large cells, but is not well adapted to support asymmetrical traffic. TDD is by definition more flexible to support traffic asymmetry, but it requires synchronization of the base stations, and is not appropriate for the large cells due to the limited guard periods between time slots. Table II gives the main characteristics of the two UMTS modes.

FDD mode is based on CDMA with a wide bandwidth (5 MHz). One of the major differences with IS95, developed by Qualcomm in the early 90s, is that no synchronization is needed between base stations, thus allowing easier deployment for operators. One of the key advantages of CDMA is its high spectral efficiency, so that UMTS operators will be able to offer, with the same spectrum, higher data rates than with GSM. When offering the same services as for GSM (voice for example), CDMA will give them more capacity per MHZ: recent evaluations have shown that the gain in terms of spectral efficiency could be in the order of 2 or 3 [5, 6].

TDD mode is based on a mix between TDMA and CDMA. Basically, the TDD frame has 15 time slots and, for each time slot, there is a possibility to support several simultaneous CDMA communications when joint detection is used.

Table II. Main characteristics of TDD and FDD modes.

Mode	FDD (Frequency domain duplex)	TDD (Time domain duplex)
Multiple access	DS-CDMA	TDMA/CDMA
Bit rate	3.84 Mchip/s	
Carrier spacing	4.4 to 5 MHz with a 200 kHz raster	
Frame length	10 ms	
Frame structure	15 time slots per frame	
Modulation	QPSK	
Spreading factor	4 to 256	1 to 16
Channel coding	Convolutional (rate 1/2 to 1/3) Turbo codes for BER < 10^{-3}	

IV. Network infrastructures

IV.1. General architecture

Figure 3 presents the general architecture of the UMTS network. It shows that UMTS is not only one new radio interface, but also a complete mobile network based on an evolution of the GSM/GPRS core network.

The UMTS core network comprises two distinct domains: circuit switched (CS) and packet switched (PS), as in GSM/GPRS networks. The core network's elements are the same: MSC (Mobile switching centre) for CS services and SGSN and GGSN for PS services. Two solutions are available to introduce UMTS: either to upgrade the existing elements or to introduce new ones supporting UMTS.

The principle of the separation between the access network and the core network through a standardized interface remains as in GSM. This new interface (Iu) is a reference point which, according to the different implementations, may correspond to one physical interface or two. However, there are always two distinct logical flows through this interface: one for the packet switched domain and the other for the circuit switched domain.

The concept of the subscriber identification module (SIM) is kept for UMTS, but with a new smart card: the UICC (UMTS integrated circuit card). This card supports a GSM SIM for GSM subscribers, the USIM for the UMTS subscribers, as well as other modules for different applications (credit cards, e-commerce, subscriptions for leisure activities).

IV.2. Access network architecture

Figure 4 represents the logical architecture of the UMTS access network. The radio network subsystem (RNS) includes the radio base stations (node B), and their controller (RNC).

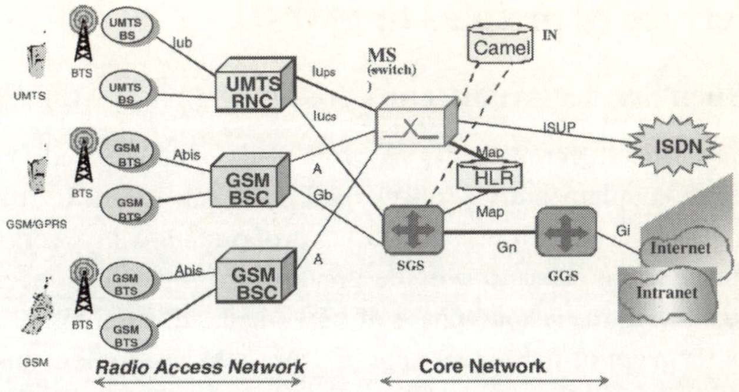

Figure 3. General architecture of the UMTS (release 99).

This hierarchical architecture, in which an entity controls several entities at a lower level, is similar to that of the GSM radio access network (BSC-BTS). Iu represents the interface between the RNC (Radio network controller) and the core network. Iub represents the interface between the nodes B and the RNC. The main difference with GSM is the existence of the Iur interface between RNCs. The main reason for the introduction of this interface is the management of macrodiversity (soft handover mechanism) in the access network. This interface will enable the management of soft handover between two node B's belonging to two separate RNCs, independently from the core network.

ATM was chosen for transport in the access network. This choice makes it possible to support all types of services (voice, circuit data, packet data, ...) that will be offered. Different AALs (ATM Adaptation Layers) will be used: AAL2 for the user data (voice or data) on the interfaces Iu-cs (circuit switched domain), Iur and Iub. AAL5 is used for signalling and the user data on the Iu-ps interface (packed-switched domain).

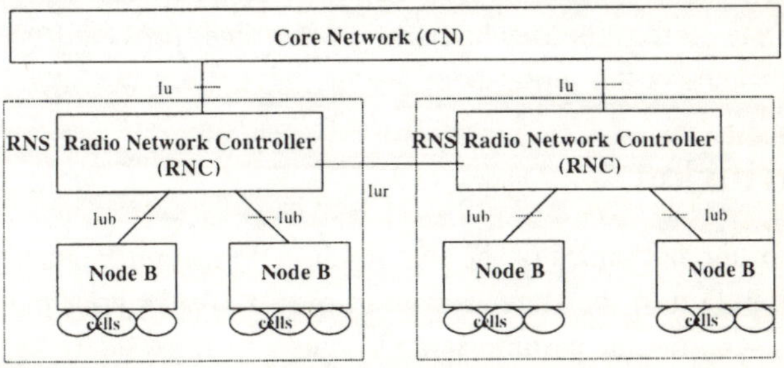

Figure 4. UMTS radio access network architecture (UTRAN).

V. Service principles in UMTS

V.1 Open Service Architecture (OSA)

For GSM, the different services were fully standardized: voice, fax, short messages, supplementary services (call hold, call forward, call conference, ...) but, for the operators, it was difficult to propose innovative services to attract the customer. So, in order to provide greater flexibility in service creation, it was decided during the second phase of GSM standardization to introduce "toolkits": CAMEL (concept of intelligent network for GSM), SIM toolkit, and MexE (Mobile Execution Environment), which includes WAP (Wireless Application Protocol). These toolkits were used in GSM to introduce prepaid services (CAMEL) or mobile internet portals (WAP). For UMTS, these principles are still valid but efforts are focused on integrating all these toolkits in a single one called OSA (Open Service Architecture). OSA is, in fact, an API based on PARLAY (PARLAY (http://www.parlay.org) is a forum developing a common API for the different networks). This new concept is still under development in 3GPP and will be introduced in the next UMTS releases.

V.2 Virtual Home Environment (VHE)

The VHE concept will enable the customer to use his services with the same ergonomics independently of his location; thus it will be possible to provide him with the same environment in his home network and when he is roaming. CAMEL (Customized Applications of Mobile network Enhanced Logic), originally developed for GSM networks, will provide roamers with the same services they use when they are in their home network.

CAMEL is based on an intelligent network architecture which separates service logic and data base from the basic switching functions, and implements the CAP (CAMEL Application Protocol) derived from INAP (Intelligent Network Application Protocol). When a subscriber is roaming, all his CAMEL data, which are stored in the HLR (Home Location Register), are transferred to the visited network. Thanks to this mechanism the service provided has the same ergonomics wherever the subscriber is.

VI. Conclusion

The choice of the principles of the UMTS radio interface in January 1998 gave a strong acceleration to the standardization process throughout the world. This decision was particularly important, because it consolidated the technical agreement between Japan and Europe on the adoption of CDMA as a common basis for UMTS. However, this was only the first step leading to the launch of UMTS networks in October 2001 in Japan and in 2002 for Europe. In 2000, most of the European countries have allocated UMTS licences using beauty contest or auctions procedures, to give sufficient time to the UMTS operators to prepare the launch of their services in 2002. A first release of the UMTS standard which is called release 99, was adopted at the beginning of 2000, and this release will be used by the manufacturers for the first generation of UMTS equipment. The competition between operators will mainly be based on their ability to provide to their customers new services because, when UMTS is launched, a high percentage of the population will have a mobile for telephony and it will be very difficult, especially for a new entrant, to attract new customers with existing services. The key aspect of UMTS will be access to high data rates and multimedia services for the customer and, without such services, it will be difficult to transform this costly adventure into success.

REFERENCES

[1] Holma (H.), Toskala (A.), WCDMA for UMTS, *John Wiley & Sons*, (2000).

[2] Mouly (M.), Pautet (M-B.), The GSM system for Mobile Communications, (1992).

[3] Blanc (P.), Charbonnier (A.), Verrier (D.), L'UMTS: la génération des mobiles multimédia, *L'écho des recherches*, n° 170, 1er trimestre, (1998).

[4] Haardt (M.), Klein (A.), Koelhn (R.), Oestreich (S.), Purat (M.), Sommer (V.), Ulrich (T.), The TD-CDMA based UTRA TDD mode, *IEEE Journal on Selected Areas in Communications*, **18**, n° 8, pp. 1375-1384, (Aug. 2000).

[5] Acx (A.G.), Mendribil (P.), Capacity evaluation of the UTRA FDD and TDD modes, *49th Vehicular Technology Conference*, Houston, **3**, pp. 1999-2003, (1999).

[6] France Telecom, Technical analysis and comparison of UTRA concepts, ETSI SMG2 Adhoc n° 4, Tdoc SMG2/UMTS 126/97, Helsinki, (17-21 Nov. 1997).

Chapter 2

Improvements in W-CDMA: principles and experimental results

M. Sawahashi, K. Higuchi, and S. Tanaka
Wireless Research Laboratories, Japan

F. Adachi
Graduate School of Engineering, Tohoku University, Japan

I. Introduction

Associated with the successful planned introduction of global commercial wideband code division multiple access (W-CDMA) [1], [2] service from this year, the dawn of the genuine era of wireless Internet is upon us. The achievable maximum information bit rate guaranteed by the required quality level in the IMT-2000 is 2 Mbps and in the near future the peak bit rate of nearly 10 Mbps will be possible for high-speed downlink packet access (HSDPA), which is now undergoing standardization in the Third Generation Partnership Project (3GPP). Therefore, rich services such as Internet access and the transmission of video and high-quality images from/to moving vehicles will be achieved in the W-CDMA system. DS-CDMA wireless access, on which W-CDMA is based, has numerous advantages over TDMA or FDMA including single frequency reuse, soft hand-off (or site diversity), enhanced radio transmission through Rake combining, and direct capacity increase through sectored antennas. The key features of the W-CDMA physical layer are:

- Inter-cell asynchronous operation and three-step fast cell search
- Flexible realization of various levels of quality of service (QoS) for various transport channels by rate matching associated with channel coding
- Signal-to-interference power ratio (SIR)-based fast transmit power control (TPC) for satisfying the required quality level for a physical channel with minimum transmit power

– Significant gains in link capacity and coverage through the use of many diversity techniques, e.g., coherent Rake time diversity using pilot symbol assisted (PSA) channel estimation, space diversity, inter-cell (sector) diversity, and transmit diversity (only in the forward link)
– High flexibility in offering different multirate services (up to 2 Mbps) through orthogonal variable spreading factor (OVSF) multiplexing and orthogonal multicode transmission
– Capacity enhanced techniques such as interference cancellation (IC) and adaptive antenna array diversity (AAAD).

The above essential W-CDMA technologies associated with its performance and the features of the W-CDMA air-interface were comprehensively overviewed in [1-3]. However, in the ongoing worldwide standardization process in the 3GPP, the radio link parameters and channel structure have been modified, and enhanced techniques such as turbo coding for high-rate data transmission and transmit diversity were adopted into the standards. Therefore, this chapter overviews the W-CDMA enhanced wireless access technologies including the channel structure and spreading code assignment in the physical layer and transport channel multiplexing into a physical channel associated with rate matching and reports on a series of laboratory and field experiments conducted in an area near Tokyo. We designed and developed an experimental system comprising a coherent multistage interference canceller (COMSIC), coherent adaptive antenna array diversity (CAAAD) receiver in the reverse link, and adaptive antenna array transmit diversity (AAA-TD) in the forward link in order to demonstrate the suppression effect on multiple access interference (MAI) and multipath interference (MPI). The experimental results of these techniques are also presented.

II. Physical channel and spreading code assignment

II.1. Physical channel [4-5]

W-CDMA has a three-layered channel structure: physical, transport, and logical. The physical channels provide several transport channels to the MAC (Medium Access Control) layer, which is a sub-layer of the data link layer (Layer 2). The MAC layer provides several different logical channels to a higher layer, that is the RLC (Radio Link Control) layer. The physical channels are classified by spreading codes, carrier frequency, and in-phase (I)/quadrature-phase (Q) assignment.

One radio frame of a physical channel has a frame length of 10 msec and comprises 15 slots. Thus, the slot length is equal to a basic updating unit of adaptive fast TPC and channel estimation of coherent Rake combining and is optimized to the value of 0.667 msec taking into account a tradeoff between frame efficiency and tracking ability of fast TPC and channel estimation against fast fading variation. The number of channel-coded information bits, which each physical channel conveys, differs according to the type of physical channel and spreading factor (SF). The features of the major physical channels are described below.

(1) P-CCPCH (Primary-Common Control Physical Channel)

One P-CCPCH is defined for each sector in the forward link. The P-CCPCH has a fixed SF of 256 (15 ksps) and carries the BCH transport channel. It is not transmitted during the first 256-chip duration, but instead the P-SCH and S-SCH are transmitted during that period at each slot.

(2) S-CCPCH (Secondary-Common Control Physical Channel)

Multiple S-CCPCHs, which are common channels in the forward link, are defined in each cell (sector) and carry paging information and lower data information from a higher layer.

(3) PRACH (Physical Random Access Channel)

Multiple PRACHs, which are common channels in the reverse link, are defined and used to carry the RACH transport channel comprising lower information data from a higher layer.

(4) DPCH (Dedicated Physical Channel)

A DPCH is assigned to each mobile station (MS) in both the forward and reverse links. It comprises a DPCCH (Dedicated Physical Control Channel) and a DPDCH (Dedicated Physical Data Channel).

A DPDCH consists of a channel-coded data sequence and more than one DPDCH can be assigned to one DPCH. A DPCCH is used for Layer 1 control of DPCH and one DPCCH is defined for one DPCH. A DPCCH comprises pilot bits for coherent channel estimation, TPC bits, TFCI (Transport Format Combination Indicator) bits, and FBI (Feedback information) bits designating the control information for transmit diversity in the forward link (thus, FBI bits are defined only in the reverse link).

(5) CPICH (Common Pilot Channel)

A CPICH is the common pilot channel used for channel estimation, path search for Rake combining (generation of power delay profile), and the third step, i.e., scrambling code identification in the three-step cell search method. Two kinds of CPICHs are defined: primary-CPICH and secondary CPICH. The primary-CPICH has two-symbol data sequences associated with two antennas. Without transmit diversity all symbol sequences with all "1" are transmitted from Antenna #1, and with transmit diversity, the second primary-CPICH with different symbol sequences from those of the first primary-CPICH are also transmitted from Antenna #2 in addition to the first primary-CPICH.

In future applications of smart antennas for spot beam transmission, the secondary-CPICH will be defined, which will be spread by the primary or secondary scrambling code.

(6) SCH (Synchronization Channel)

The SCH is a common channel in the forward link, which is used for cell search. Primary and secondary-SCHs are used for the first step and second step for the three-step cell search method. They are transmitted only during the 256-chip period at the beginning of each slot.

(7) AICH (Acquisition Indication Channel)

The AICH is a common channel in the forward link used for random access control. It is used as a pair comprising a PRACH and PCPCH.

(8) PICH (Page Indication Channel)

The PICH is a common channel in the forward link and is associated with S-CCPCH, in which the PCH transport channel is mapped.

(9) PDSCH (Physical Down Link Shared Channel)

The PDSCH is a common channel in the forward link, which carries the DSCH transport channel and is used for high rate packet data transmission.

(10) PCPCH (Physical Common Packet Channel)

The PCPCH is a common channel in the reverse link, which carries the CPCH transport channel and is used for high rate packet data transmission.

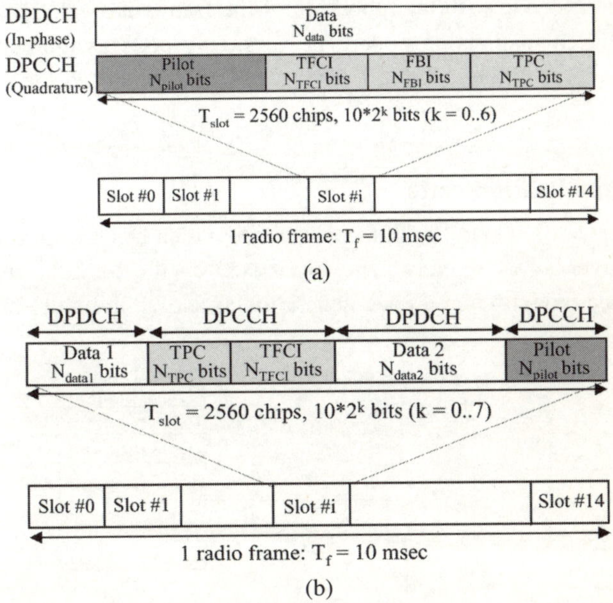

***Figure 1**. Frame structure of DPCH (a) reverse link (b) forward link.*

The frame structure of the DPCH in the reverse and forward links is illustrated in Figures 1(a) and 1(b), respectively. The DPDCH and DPCCH are code-multiplexed into I and Q channels, respectively, in the reverse link. Since the DPCCH with a fixed rate (SF) and DPDCH with variable date transmission are separated from each other in the orthogonal phase, fluctuation of the amplitude during variable transmission can be decreased. Meanwhile, the DPCCH and DPDH are alternatively time-multiplexed within a slot in the forward link.

Table I. *Spreading code assignment.*

		Channelization code		Scrambling code
		Repetition period = Data symbol period		Repetition period = 10 msec frame
Forward link		User identification (4-512 chips)		Cell (Sector) identification (38,400 chips)
	CPICH	#0	SF = 256	Primary
	P-CCPCH	#1	SF = 256	Primary
	S-CCPCH	Arbitrary	SF = 4-256	Primary (Secondary)
	DPCH	Arbitrary	SF = 4-256	Primary (Secondary)
	AICH	Arbitrary	SF = 256	Primary (Secondary)
	PICH	Arbitrary	SF = 256	Primary (Secondary)
Reverse link		Code-channel identification in multicode transmission (4-256 chips)		User identification (38,400 chips)
	DPCH	Arbitrary	SF = 4-256	Primary (Secondary)

II.2 Spreading code assignment [6]

W-CDMA adopts a two-layered spreading code assignment, which combines a channelization code with the repetition period of the corresponding symbol rate and a scrambling code with the repetition of the frame interval. The OVSF code is used as the channelization code. The spreading code assignment for each physical channel is given in Table I. The SF of 4 to 256 is used for S-CCPCH and DPCH.

II.2.1. Channelization code

Starting from $C_{ch,1,0}$ (1) (SF = 1), the OVSF code which has a length of 2^{k-1}-chip at the *k-th* layer, is recursively generated based on the formula given below, resulting in the tree-structured code generation as shown in Figure 2 [7].

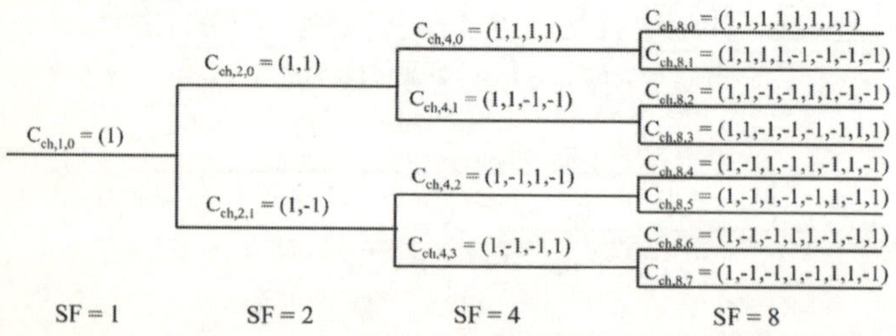

Figure 2. *Generation method of OVSF codes.*

$$(1)\quad \begin{bmatrix} C_{ch,2(k+1),0} \\ C_{ch,2(k+1),1} \\ C_{ch,2(k+1),2} \\ C_{ch,2(k+1),3} \\ \vdots \\ C_{ch,2(k+1),2(k+1)-2} \\ C_{ch,2(k+1),2(k+1)-1} \end{bmatrix} = \begin{bmatrix} C_{ch,2^k,0} & C_{ch,2^k,0} \\ C_{ch,2^k,0} & -C_{ch,2^k,0} \\ C_{ch,2^k,1} & C_{ch,2^k,1} \\ C_{ch,2^k,2} & -C_{ch,2^k,1} \\ \vdots & \vdots \\ C_{ch,\,2^k,2^k-1} & C_{ch,\,2^k,2^k-1} \\ C_{ch,\,2^k,2^k-1} & C_{ch,\,2^k,2-1} \end{bmatrix}$$

The k OVSF codes of the *k-th* layer are orthogonal to each other. Furthermore, any two codes belonging to different layers are orthogonal except for when one code is not the mother code of the other. For example, $C_{ch,2,0}$ and $C_{ch,4,2}$ are orthogonal to each other. When $C_{ch,2,0}$ is already assigned, any code below this code on the code tree cannot be used, this is a restriction of the code assignment. The codes of $C_{ch,256,0}$ and $C_{ch,256,1}$ are commonly used for all cells for the P-CPICH and P-CCPCH in the forward link, respectively. The channelization codes of other physical channels are assigned from a higher layer.

II.2.2. Scrambling code

Cell (sector)-specific and user-specific scrambling codes are assigned in the forward and reverse links, respectively. In the reverse link, the repetition period of the scrambling code is 10 msec and that with the repetition period of 256 chips is optionally defined for future application of multiuser detection. The long scrambling code is truncated by a duration of 38,400 chips from the beginning of the Gold sequence with the repetition period of 2^{24} chips. There are 2^{24} long scrambling codes.

The scrambling code in the forward link is generated by truncating the 38,400 chips from the beginning of the Gold sequence with the repetition period of 2^{18} and its shifted version by 131,072 chips. The 8,192 scrambling codes are grouped into 512 scrambling-code groups, where each group comprises 1 primary scrambling code with 15 corresponding secondary scrambling codes. The primary scrambling code is first used, and then the secondary scrambling codes are used to cover any shortage in the channelization code set associated with the primary scrambling code.

Five hundred twelve primary scrambling codes are divided into 64 primary-scrambling-code groups (hereafter we simply denote group), each including 8 primary scrambling codes. This group-wise divided primary scrambling code structure is used for the three-step cell search algorithm, which is described in Section III.

II.2.3. Synchronization code

A synchronization code is used to spread a SCH and comprises a primary synchronization code (PSC) and secondary synchronization code (SSC) both with the length of 256 chips, which are used for P-SCH and S-SCH, respectively. Let PSC be denoted as C_{PSC}, in which C_{PSC} is a complex-value code sequence with the same sequence for real and imaginary parts expressed as

(2) $C_{PSC} = (1 + j) \times \{a, a, a, -a, -a, a, -a, -a, a, a, a, -a, a, -a, a, a\}$, where

$$a = \{x_1, x_2, x_3, \ldots, x_{16}\} = \{1, 1, 1, 1, 1, 1, -1, -1, 1, -1, 1, -1, 1, -1, -1, 1\}$$

Let 16 SSCs be denoted as $C_{SSC,k}$ (k = 1, 2, ..., 16). Then, $C_{SSC,k}$ is generated by multiplying the *j-th* component ($1 \leq j \leq 256$) of vector Z of a common sequence with the length 256 chips and the *j-th* component of the *n-th* column of H_8 of the Hadamard matrix, where $n = 16 \times (k - 1)$. Let $h_n(j)$ and $z(j)$ be the *j-th* symbol of *n-th* column of the Hadamard matrix and the *j-th* symbol of a common sequence, respectively. By selecting 16 columns from 256 columns every 16 columns, the 16 $C_{SSC,k}$ is generated as

(3) $C_{SSC,k} = (1 + j) \times \langle h_n(0) \times z(0), h_n(1) \times z(1), h_n(2) \times z(2), \ldots, h_n(255) \times z(255) \rangle$,

where $Z = \{b, b, b, -b, b, b, -b, -b, b, -b, b, -b, -b, -b, -b, -b\}$,
and $b = \{x_1, x_2, x_3, x_4, x_5, x_6, x_7, x_8, -x_9, -x_{10}, -x_{11}, -x_{12}, -x_{13}, -x_{14}, -x_{15}, -x_{16}\}$.

II.2.4. Spreading

In the reverse kink, the channelization code is independently spread into I/Q channels by using different OVSH codes and weighted by weighting factor G, which denotes the transmitted amplitude (power) ratio of DPCCH to DPDCH. Complex spreading is applied to the physical channel: one is a code truncated by 38,400 chips from the beginning of the Gold sequence with the repetition period of 2^{24}, and the other is truncated by 38,400 chips of the shifted first Gold sequence by 16,777,233 chips. Thus, the spreading using channelization codes and the scrambling codes are expressed as

(4) $S_I = D_I C_I - D_Q C_Q$

$$S_Q = D_I C_Q + D_Q C_I$$

where $D_{I(Q)}$ denotes the I/Q components of the chip data sequence spread by channelization codes and $C_{I(Q)}$ represents the I/Q components of a long scrambling code. In this QPSK spreading, the carrier phase transition by π-degrees occurs across the zero point, thus incurring increasing nonlinear distortion of the power amplifier. Therefore, in the 3GPP standard, the HPSK (hybrid PSK) scheme was adopted, which decreased the possibility of the phase transmission crossing the zero point [6, 51]. The long scrambling codes sequence used for spreading are generated from the two original scrambling codes based on the following equation:

(5) $C_{long,\,n}(i) = C_{long,1,\,n}(i)(1 + j\,(-1)^i C_{long,2,\,n}(2\lfloor i/2 \rfloor))$

In the forward link, P-SCH and S-SCH are spread by only primary and secondary synchronization codes, respectively, commonly used for both I/Q channels. The other physical channels except for SCH are first spread by an identical channelization code with SF = m for both the I/Q channels and then complex-scrambled by the two scrambling code sequences.

III. Transport channel multiplexing

III.1. Explanation of data format for layer 1 [8]

We first explain the terminology used for data transfer between the MAC layer and Layer 1. A transport block, which corresponds to a RLC (Radio Link Control)-PDU (Protocol Data Unit), is a basic unit for data transfer between the MAC layer and Layer 1. Cyclic redundancy check (CRC) for error detection in Layer 1 is added to every transport block. One example of a transport block transfer between the MAC layer and Layer 1 is illustrated in Figure 3. A set of transport blocks simultaneously transferred between the MAC layer and Layer 1 on the same transport channel is called a transport block set. The size of the transport block is the length of the transport block defined in bit form. The size of each transport block belonging to one transport block set is uniform and is a fixed value. The number of bits within a transport block set is called the transport block set size. As shown in Figure 3, the arrival time interval of transport block sets between the MAC layer and Layer 1 is called the transmission time interval (TTI), which is equal to the channel interleaving length. The TTI is some integer

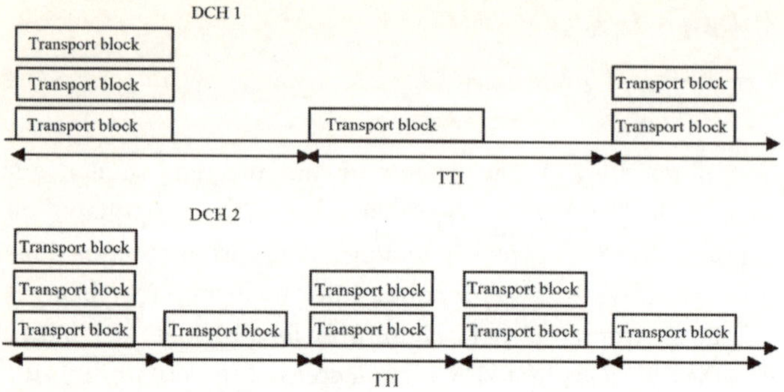

***Figure 3**. Example of exchange of data between a MAC layer and Layer 1.*

times the radio frame length (= 10 msec) and is defined as 10, 20, 40, or 80 msec in the 3GPP. The transport format is a format in which a transport clock set is transferred between the MAC layer and Layer 1 on a transport channel every TTI. The transport format comprises two attributes: the dynamic part and semi-static part. Attributes of the dynamic part are the transport block size, transport block set size, and TTI, and those for the semi-static part are error of the correction scheme such as the type of error correction and coding rate and the size of the CRC. The transport format set (TFS hereafter) is defined as a set of transport formats used for the transport channels. Within one TFS, the semi-static parts of all transport channels are identical; however, the dynamic parts may be changed every TTI in order to achieve variable rate transmission. The transport channels are simultaneously multiplexed into Layer 1 as a coded composite transport channel (CCTrCH). Each transport channel in the CCTrCH has an available TFS; however, only one transport format is used at each TTI. Thus, the combination of possible transport formats of all transport channels transferred on the same Layer 1 at each TTI is defined as a transport format combination (TFC). Furthermore, a set of TFC applied to the CCTrCH is called as transport format combination set (TFCS). The indicator designating the TFC I called the transport format combination indicator (TFCI). TFCI bits are multiplexed into the DPCCH of each DPCH. In the receiver, the TFCI bits are used to decode Layer 1 data sequences and de-multiplex transport blocks transferred on one physical channel. In addition to the explicit TFCI detection method, the blind transport format detection method using CRC to trace the surviving trellis path ending at the zero state among the possible transport formats is also specified in the 3GPP standard (note that blind detection is used only for the forward link) [9].

III.2. Transport channel [4, 8]

A transport channel is defined as a channel that is used to transfer various kinds of data to the MAC layer. The major transport channels are described below. The mapping relationships between the major physical channels and transport channels are given in Figure 4.

(1) BCH (Broadcast Channel)

The BCH is a forward link transport channel that is used for broadcasting system – and cell-specific information. The BCH is always transmitted over the entire cell and has a single transport format.

(2) FACH (Forward Access Channel)

The FACH is a forward link transport channel that is commonly used for multiple MSs and for transmitting low-rate user information from a higher layer.

(3) PCH (Paging Channel)

The PCH is a forward link transport channel that is transmitted over the entire cell and is used to transmit paging information.

(4) RACH (Random Access Channel)

The RACH is a reverse link transport channel, which is received from the entire cell. The RACH is characterized by collision risk and by being transmitted using open-loop transmit power control.

(5) DCH (Dedicated Channel)

The DCH is a forward link and reverse link transport channel, which is transmitted over the entire cell or only a part of the cell using a smart antenna. The DPCH is used for the transmission of user data and is assigned to each MS. Variable rate transmission and fast transmit power control (TPC) are applied to the DPCH.

(6) DSCH (Down Link Shared Channel)

The DSCH is a forward link transport channel shared by several MSs. The DSCH is used for mainly high-rate packet data transmission and is transmitted over the entire cell or over only a part of the cell using beam-forming antennas.

(7) CPCH (Common Packet Channel)

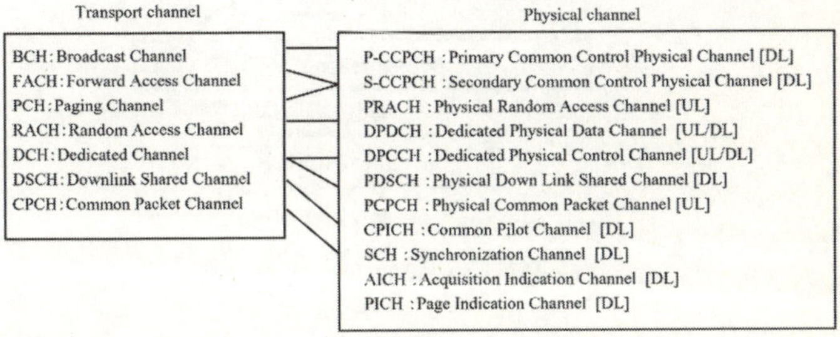

***Figure 4**. Relation between physical channel and transport channel.*

The CPCH is a reverse link transport channel and is associated with a dedicated channel on the forward link, which provides power control and CPCH control commands. The CPCH is used for high-rate data transmission on random access channels.

III.3. Multiplexing and rate matching [9]

The flow of the transport channel multiplexed into a physical channel in the reverse link is depicted in Figure 5. First, CRC parity bits required for block error detection at the receiver are calculated for the original data sequence per transport block of each transport channel. Then, the calculated CRC bits are attached to each transport block. All transport blocks with CRC bits within one TTI are serially concatenated followed by channel coding. For channel coding, convolutional coding or turbo coding are used in the 3GPP specification. For the common transport channels such as BCH, PCH, and RACH, convolutional coding with the rate of 1/2 and the constraint length of 9 bits is used. Convolutional coding with the rate 1/3 (1/2) is also used for FACH and DPCH with a lower channel bit rate, and turbo coding [10] with the rate 1/3 and the constraint length of 4 bits is used for FACH and DPCH with higher channel bit rates. After the coded data sequence of each transport channel is interleaved over the length of the TTI

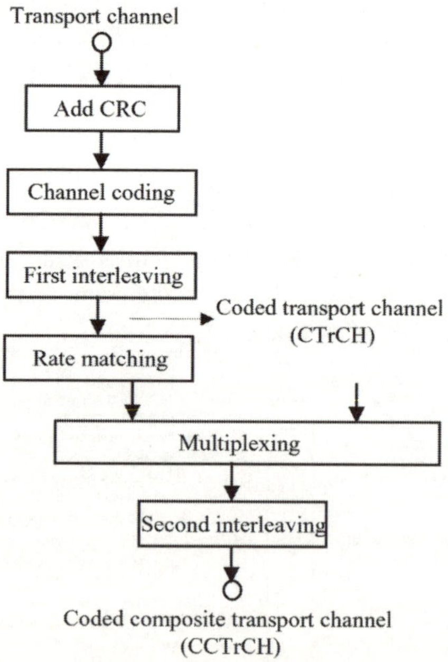

Figure 5. Transport channel multiplexing (reverse link).

(first interleaving), rate matching is performed according to the required QoS and the number of bits. The data sequence of each transport channel after rate matching is segmented and interleaved over one radio frame length (second interleaving). Finally, the CCTrCH containing all transport channels is multiplexed into a physical channel. As described previously, the first channel interleaving is performed before rate matching of each transport channel in the reverse link. Meanwhile, discontinuous transmission (DTX) is allowed when there is no transmitted data sequence in the radio frame of a certain transport channel in the forward link. Thus, the rate matching is performed independently for each transport channel before the first interleaving.

As shown in Figure 6, transport channels with different bit rates and QoS levels are multiplexed and transferred into one physical channel. A transport block is a basic unit for data transfer between the MAC layer and Layer 1 (in Figure 6 of the transport channel, 1(1) represents the first block of transport channel 1). The required QoS, i.e., the block error rate (BLER) or bit error rate (BER) of the physical channel is achieved by changing the transmit power or data modulation scheme according to the fading variation. In general, the QoS level of one physical channel can be controlled by changing the target SIR of fast TPC using outer loop control so that the output BLER or BER is equal to the required value as explained later. However, the average received signal energy per bit-to-interference and background noise spectrum density (E_b/I_0), thus, the received signal power, is an almost constant value during one radio frame interval. Therefore, in order to bundle various transport channels with different QoSs into one physical channel, the required QoSs of various transport channels are simultaneously satisfied with respect to the identical average received signal power by changing the number of coded bits of each transport channel after channel decoding (this process is called rate matching). That is to say, by repeatedly transmitting some coded bits at a regular interval, the BLER or BER is

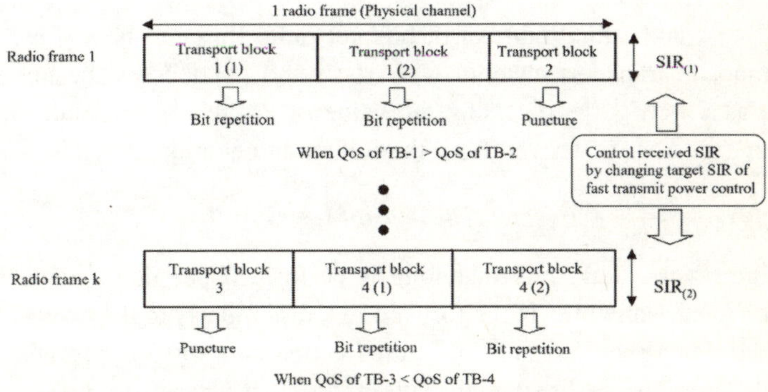

***Figure 6**. Principle of rate matching.*

improved. Contrarily, if encoded bit sequences are punctured at a regular interval, the received quality is degraded. In this way, the number of bits of each transport channel multiplexed into the physical channel is flexibly changed every radio frame by rate matching described hereafter.

In the reverse link, rate matching is performed for the coded data sequence of each transport channel after the first interleaving. The number of bits of each transport channel to be repeated or punctured is calculated based on the rate-matching attributes signaled from a higher layer. The DTX, when there is no coded transmitted data sequence of a certain transport channel multiplexed into a physical channel, is not permitted. Thus, the spreading factor (SF), i.e., equivalently the symbol date rate, of a physical channel is first determined according to the total number of bits per radio frame of all transport channels multiplexed into the physical channel. Then, rate matching is performed so that the sum of the bits of all transport channels per radio frame after rate matching should equal the bits per radio frame accommodated into the physical channel having the assigned SF. Let $N_{i,j}$ and $\Delta N_{i,j}$ be the number of coded data bits of transport channel *i* per radio frame with TFC j before rate matching and the number of bits per radio frame to be bit-repeated or punctured (the positive and negative values of Δ denote the bit-repetition and puncture), respectively. The value of $Z_{i,j}$ which is needed for the calculation of $\Delta N_{i,j}$ is recursively computed from the following equations using the rate-matching attribute value, RM_i.

$$Z_{0,j} = 0$$

$$(6) \qquad Z_{i,j} = \left\lfloor \frac{\left\{\left(\sum_{m=1}^{i} RM_m \times N_{m,j}\right) \times N_{data,j}\right\}}{\sum_{m=1}^{I} RM_m \times N_{m,j}} \right\rfloor \qquad \textit{for all } i = 1, \ldots, I$$

where $N_{data,j}$ is the total number of bits per radio frame to be assigned to code the composite transport channel with TFC j and $\lfloor x \rfloor$ denotes the integer value defined as $x - 1 \le \lfloor x \rfloor \le x$. Using the value of $Z_{i,j}$ recursively calculated from Equation (6), $\Delta N_{i,j}$ is derived from the following equation.

$$(7) \qquad \Delta N_{i,j} = Z_{i,j} - Z_{i-1,j} - N_{i,j} \qquad \textit{for all } i = 1, \ldots, I.$$

In the reverse link, rate matching is performed per radio frame based on Equation (7). Meanwhile, in the forward link dissimilarly to the reverse link, DTX is applied when there are no transmitted coded data bits of a certain transport channel. Thus, the rate-matching pattern does not necessarily change for each radio frame. Rate matching is performed as follows. The number of bits per TTI of transport channel i before rate matching, $N_{i,h}^{TTI}$, is first calculated for the corresponding TFC h belonging to TFCS. Then, from the value of $N_{i,h}^{TTI}$, and the number of radio frames of transport channel i over TTI, F_i, the corresponding number of bits per radio frame was derived for all TFC belonging to TFCS. Thus, rate matching is performed such that the number of total bits per radio frame for TFC h_{Max}, when the summation of bits per radio frame of all transport channels is maximized, is equal to the number of bits per radio frame accommodated into a physical channel, that is to say, the number of bits per radio frame. Then the number of bits per TTI to be bit-repeated or punctured is computed for each transport channel. Based on this obtained rate matching pattern, the number of bits per radio frame of each transport channel is updated every TTI. Consequently, when transport channels having different TTI are multiplexed, the number of total bits belonging to a radio frame is changed at the shortest TTI at every TTI. If the number of bits per radio frame of transport channel i after rate matching is lower than the maximum number of bits assigned to that transport channel, DTX is performed during an interval corresponding to the number of bits to be shortened.

IV. Asynchronous cell sites and three-step search method

In asynchronous cell site operation, which is the most prominent feature in W-CDMA, flexible system deployment from outdoors to indoors is possible, since no external timing source such as the global positioning system (GPS) is required. To allow asynchronous cell site operation, two-layer spreading code allocation is used [1]. In the forward link, cell sites are distinguished by their unique scrambling codes, and data channels (control and traffic channels) in each cell site are distinguished by different OVSF codes. To reduce the cell search time in asynchronous cell site operation, we proposed a three-step cell search method using scrambling code masking [11]. Subsequently, our original cell search method was refined in the standardization process. The forward link frame structure in the 3GPP standard required for the three-step cell search is illustrated in Figure 7. The base station (BS) transmits a continuous common pilot channel (CPICH), primary synchronization channel (SCH), and secondary-SCH over the 256-chip duration at the beginning of each slot (every 0.667 msec). The spreading codes for the CPICH and the DPCHs are taken from a set of OVSF codes, thereby maintaining mutual orthogonality between the CPICH and DPCHs. These channels are further scrambled by a cell-specific scrambling code with a 10-msec repetition period (= 38,400-chip duration), which is equal to the data frame

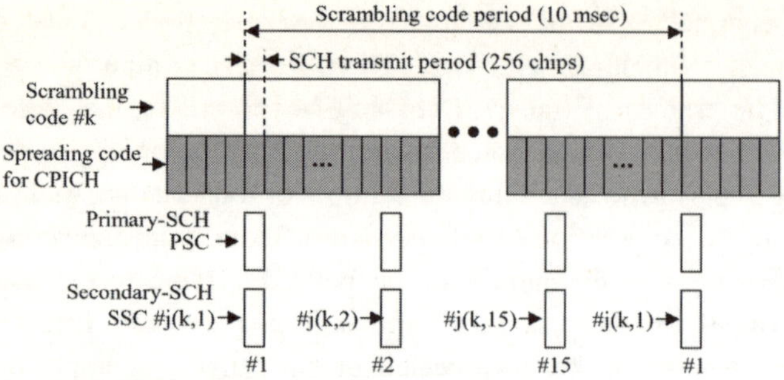

***Figure 7**. Forward link frame structure of CPICH and SCH.*

length. The PSC for the primary-SCH is common to all cell sites and the SSC for the secondary-SCH denotes the group index into each of which the scrambling codes are grouped beforehand. The total number of scrambling codes to be searched is 512, which is divided into 64 groups of 8 codes each. The transmit powers of the primary- and secondary-SCH are set to half that of the CPICH.

The operational flow of the three-step cell search algorithm is illustrated in Figure 8. Using SCHs and CPICH, the three-step cell search is performed as follows. First, the PSC-matched filter (MF) is used. The MF output is averaged over period T_1 to detect the primary-SCH time position that provides the maximum average correlation. Next, the scrambling code group is identified by taking the

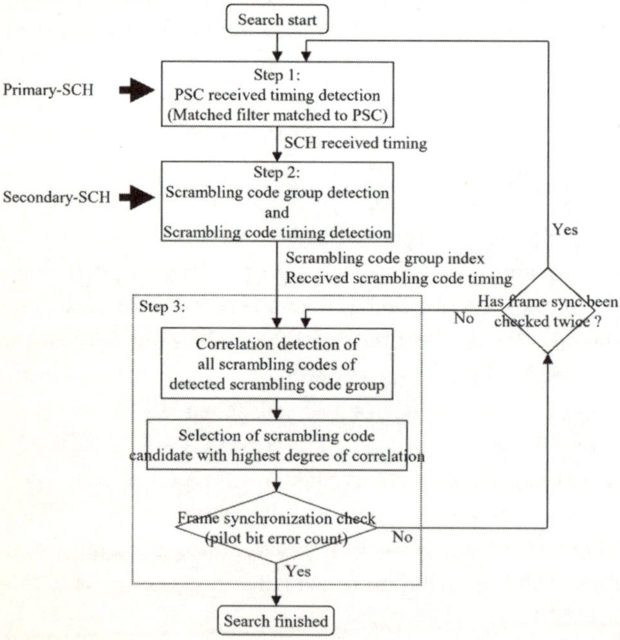

***Figure 8**. Operational flow of three-step cell search method.*

cross-correlation between the received signal and the set of SSCs over period T_2. Finally, the scrambling code is identified by taking a partial correlation between the received signal and each of the candidate scrambling codes and then averaging over period T_3. The scrambling code that provides the maximum correlation is determined as the scrambling code to be searched. To reduce false detection, a verification mode is added by using a frame synchronization check. When the synchronization verification failed two consecutive times, the cell search process is restarted from the first step. The correlation peaks of PSC and SSC calculated in the first and second steps are averaged during T_1 and T_2 in order to reduce the influence of MAI and the background noise components. However, especially when the velocity of a MS is low, the probability for false detection in the first and second steps is greater since the duration of low received signal power due to fading becomes longer. Thus, time space transmit diversity (TSTD) is applied to SC in the 3GPP specification, with which primary- and secondary-SC are alternatively transmitted slot-by-slot from different antennas [5]. Since a successive primary- and secondary-SCH are transmitted from different antennas having a low fading correlation, the false detection is decreased due to the transmit diversity effect.

Figure 9 shows the measured laboratory experimental results of the probability distribution of the cell search time with the fading maximum Doppler frequency, f_D, as a parameter using the 4.096 Mcps WCDMA experimental system with TSTD [12, 13]. In addition to CPICH and SCHs, 10 DPCHs without fast TPC were transmitted as a channel load. An L = 2 path Rayleigh

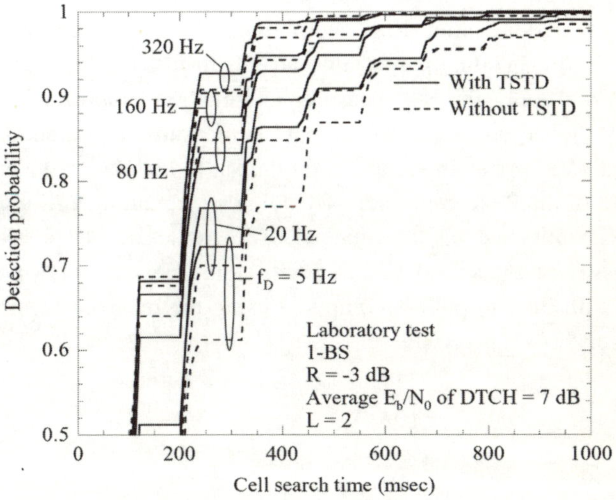

Figure 9. *Probability of distribution of cell search time using TSTD.*

fading channel with average equal power was assumed because we confirmed that field experimental results conducted near Tokyo could be well approximated using this model where 2 – 3 paths with unequal average received signal power were observed. The transmit power ratio of CPICH to DPCH and average received E_b/N_0 of DPCH were set to – 3 dB and 7 dB, respectively. We set T_1, T_2, and T_3 to 40, 30, and 10 msec, respectively. Figure 9 shows that as f_D becomes larger, the cell search time becomes shorter since false detection is decreased. The figure also shows that by using TSTD, the cell search time when f_D is low such as 5 and 20 Hz can be decreased because false detection is mitigated when the received signal level drops. As a result, the cell search time at the detection probability of 90% with TSTD is decreased by approximately 100 msec compared to that without TSTD. The cell search can be completed within approximately 250 msec at the probability of 90% with TSTD, when R = – 3 dB and f_D = 5 Hz.

V. SIR measurement-based fast TPC

Fast TPC based on SIR measurement of Rake combined signals is used to minimize always the transmit power according to the traffic load both in the reverse and forward links. This results in increased capacity by reducing the interference to other users in other cells and the user's own-cell. Fast TPC comprises two loops as shown in Figure 10: the inner loop and outer loop.

Inner loop operation is performed as follows. In the Rake combiner, the despread signals associated with resolved paths are multiplied by the complex conjugate of their channel gain estimates and summed. Therefore, if the SIR measurement is done after Rake combining, it is affected by the channel estimation error. In this paper, instead of measuring the SIR after Rake combining, we apply the SIR measurement method proposed in [14, 15], in which, first, the SIR on each resolved path is measured and then, the SIRs of all the resolved paths are summed to obtain the SIR (which is equivalent to the one at the output of the Rake combiner). By doing so, obtaining an SIR measurement that has less influence on the channel estimation error is possible. The SIR measurement is summarized below. First, signal power $\tilde{S}_l(k)$ of the *k-th* slot associated with the *l-th* path is computed using the received N_p pilot symbols. Signal power $\tilde{S}_l(k)$ is given by

(8) $$\tilde{S}_l(k) = |\bar{r}_l(k)|^2$$

where

(9) $$(\bar{r}_l(k) = \frac{1}{N_p} \sum_{n=0}^{N_p-1} r_l(n, k) \exp\left(-\frac{j\pi}{4}\right),$$

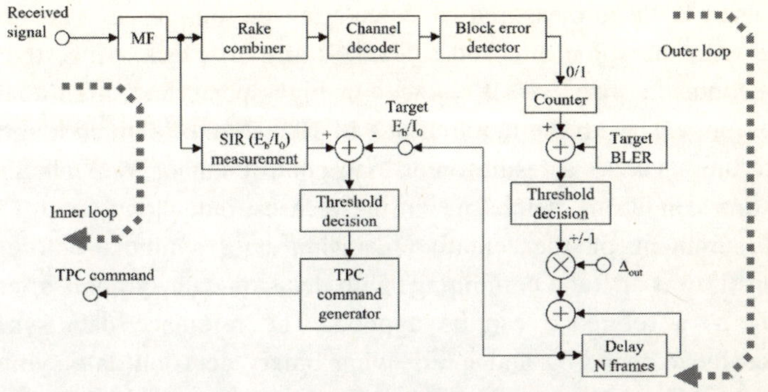

***Figure 10**. SIR-based adaptive TPC with outer loop control.*

since we assume that the modulation phase of N_p pilot symbols is $\pi/4$ radians. The instantaneous interference plus background noise power of the *l-th* path, $\tilde{I}_l(k)$, is computed as the squared error of the received N_p pilot samples

$$(10) \quad \tilde{I}_l(k) = \frac{1}{N_p} \sum_{n=0}^{N_p-1} \left| r_l(n, k) \exp\left(-j\pi/4\right) - \bar{r}_l(k) \right|^2$$

Then, $\tilde{I}_l(k)$ is averaged using a first order filter with forgetting factor $\mu(< 1)$ to obtain

$$(11) \quad \bar{I}_l(k) = \mu \bar{I}_l(k-1) + (1 + \mu)\, \tilde{I}_l(k)$$

The SIR at the *k-th* slot associated with the *l-th* path $\tilde{\lambda}_l(k)$ is given by

$$(12) \quad \tilde{\lambda}_l(k) = \tilde{S}_l(k)/\bar{I}_l(k)$$

Finally, the SIR at the *k-th* slot, $\bar{\lambda}(k)$, is obtained as

$$(13) \quad \bar{\lambda}(k) = \sum_{l=0}^{L-1} \tilde{\lambda}_l(k)$$

The measured SIR was compared to the target SIR and the TPC command was generated, which was transmitted to raise or lower the mobile transmit power by ± 1 dB every 0.667 msec. Even if the received SIRs are the same, the received quality (BLER) is not the same because the BLERs are affected by the number of paths, maximum Doppler frequency (which depends on the speed of the vehicle), and SIR measurement, etc. Therefore, the outer loop controls the target SIR with a more gradual updating interval compared to the inner loop so that the measured BLER or BER is equal to the target value. In general, a BLER-based outer loop is used. BLER is measured by calculating the number of CRC results that coincide with the value attached to every transport block. Since the required BLER becomes a very small value for high-speed and high-quality data transmission, e.g., with the required BER of 10^{-6}, it takes a much longer time to calculate the BLER. As a result, outer loop control cannot track changes in the propagation conditions. Therefore, in these cases, outer loop control based on BER measurement of the tentative decision data symbols before channel decoding (i.e., after Rake combining) with decision data symbols after channel decoding as a reference can be applied. The reference data symbols are generated by re-encoding and interleaving binary decision data symbols after channel decoding. Although data decision error occurs in the decoded data sequence, it is considered that the impact on the reference symbols is very small.

VI. Diversity

VI.1. Coherent Rake combining (Rake time diversity)

PSA coherent detection is used for both the reverse and forward links [16, 17]. The block diagram of the PSA coherent Rake combiner is illustrated in Figure 11(a). The received multipath signals are despread by the MF and resolved into L-multipath components of transmitted quaternary phase shift keyed (QPSK) modulated data that are received via different propagation paths with different delay times. The coherent Rake combiner output is expressed at the *n-th* symbol position of the *k-th* slot associated with the *l-th* path (l = 0,1, ..., L –1) using despread signal $r_l(n, k)$, as

$$(14) \qquad \tilde{d}(n, k) = \sum_{l=0}^{L-1} r_l(n, k)\,\tilde{\xi}_l^*(k),$$

where $\tilde{\xi}_l(k)$ represents the channel estimates. The output data sequence, $\tilde{d}(n, k)$, is de-interleaved and channel decoded to recover the transmitted binary data sequence. In order to achieve accurate channel estimation that works satisfactorily in a fast fading environment, we presented an improved channel estimation filter called a weighted multislot averaging (WMSA) channel estimation filter [17] as shown in Figure 11(b). After obtaining the instantaneous channel estimates of each slot, the channel estimates, $\hat{\xi}_l(j + i)$s, of 2J-multiple slots (i = – J + 1, ..., 0, 1, ..., J) are then weighted and summed to obtain the final channel estimate, $\xi_l(k)$, as

$$(15) \qquad \tilde{\xi}_l(k) = + \sum_{i=0}^{J-1} \alpha_i \hat{\xi}_l(k-i) + \sum_{i=1}^{J} \alpha_i - 1\,\hat{\xi}_l(k+i)$$

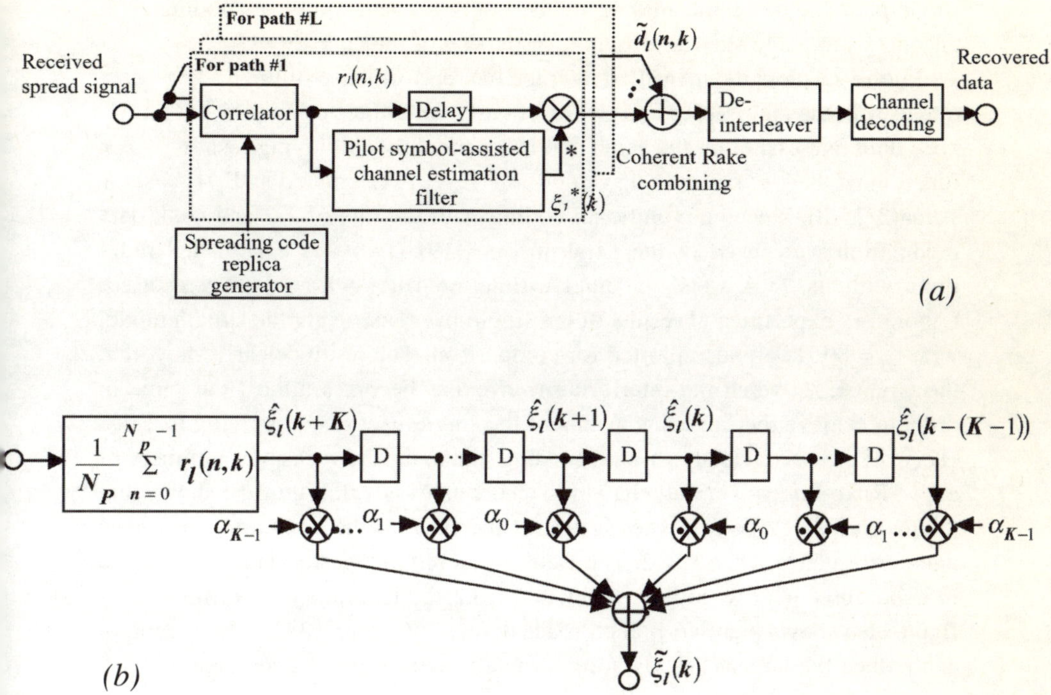

Figure 11. Coherent Rake receiver. (a) Receiver structure (b) WWSA channel estimation filter.

where α_i is the real-valued weight. Using the WMSA channel estimation filter, accurate channel estimation is possible, particularly in slow fading environments. The optimum value of α_i varies according to the fading correlation between succeeding slots in a real fading channel. Therefore, we proposed in [18] an adaptive WMSA channel estimation filter, in which a weighting factor is adaptively controlled by measuring an inner product of the averaged despread pilot signals of successive slots.

We evaluated the BER performance of coherent Rake combining with SIR based fast TPC in field experiments conducted in an area near Tokyo. The cell site and mobile transmitter/receiver antennas were located 59 and 2.9 m off the ground, respectively. A measurement vehicle equipped with the mobile receiver was driven along roads at distances of 0.75 – 1.35 km from the cell site at the average speed of approximately 30 km/h. The measurement course passes through a business zone, lined with office buildings and factories. Other conditions are given in detail in [19]. The average delay spread of the test course was approximately 1 μsec. The test course first experienced clear two-path and single-path fading at the middle of the course. Then, three-path fading with unequal average power was observed at the end of the course.

Figure 12 plots the measured average BER performance of the 32-kbps data rate user in the single-user and two-user cases (one-interfering user with a 64-kbps data rate assuming the same BER independently employing fast TPC), as a function of the TPC target E_b/I_0 value (note E_b/I_0 is calculated as E_b/I_0 = SIR + 10log(3/2) dB, since convolutional coding with the rate of 1/3 and QPSK data modulation were used in the experiments) [19]. Two MSs established radio links with BS 1. A WMSA channel estimation filter with J = 2 was used. Laboratory experimental results of the single-user case using the L-path model with f_D = 80 Hz are also plotted for comparison. The results clearly show that the target E_b/I_0 when the interfering user exists becomes almost the same in order to achieve the same BER as that of the single-user case, implying that fast TPC worked satisfactorily in a real fading channel. The measured numbers of active Rake fingers per antenna along test courses is 2.0. Figure 12 shows that the measured BER performance is almost the same as the laboratory-measured BER performance when L = 2. The field-measured BER performance results are in good agreement with those estimated from the laboratory experiment. The figure also shows that two-branch space diversity (antenna diversity) reception can reduce the target E_b/I_0 by approximately 3 dB at the average BER of 10^{-3}.

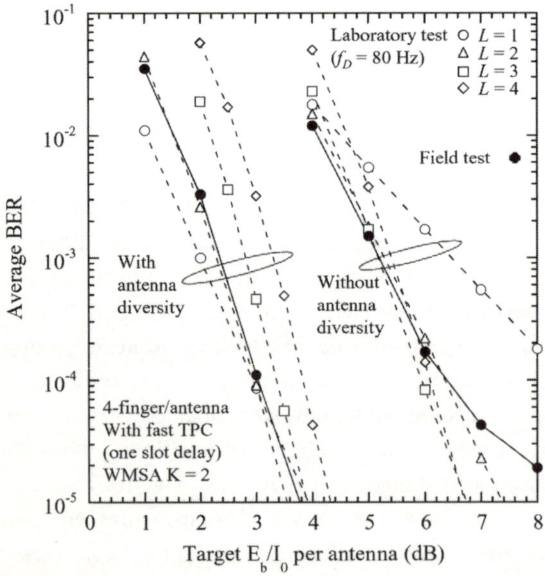

Figure 12. Average BER as a function of target E_b/I_0 per antenna. Field experiments.

With space diversity reception, the average BER of 10^{-3} can be achieved at the required E_b/I_0 of approximately 3 dB per antenna.

The measured average BER performance of the 64 kbps channel using turbo coding is plotted in Figure 13 as a function of the MS relative transmit power with the channel interleaving length of T_{CHL} = 40 msec [20]. Turbo coding with the rate of R = 1/3 and the constraint length of K = 4 bits (generator polynomials are 13, 15, and 15 in octal notation) were used, while the rate and constraint of convolutional coding as a reference were R = 1/3 and K = 9 bits, respectively. Primary interleaving (PIL) [9, 22] and multistage interleaving (MIL) [9, 21], which offer a greater capability for randomization compared to the block interleaving method, were used as turbo interleaving and channel interleaving methods, respectively. In the experimental system, Max-log – Map decoding was used as the soft-in/soft-out decoder and the number of iterations, m, was assumed to be eight, which was sufficiently large. From Figure 13, the MS average transmit power for the average BER of 10^{-6} using turbo coding can be decreased by approximately 0.6 (0.3) dB compared to that using convolutional coding without (with) antenna diversity reception. Although the superiority of turbo coding to convolutional coding was confirmed in an actual multipath fading channel, this difference was decreased compared to the laboratory experiments assuming a fixed delay time for each path using a fading simulator,

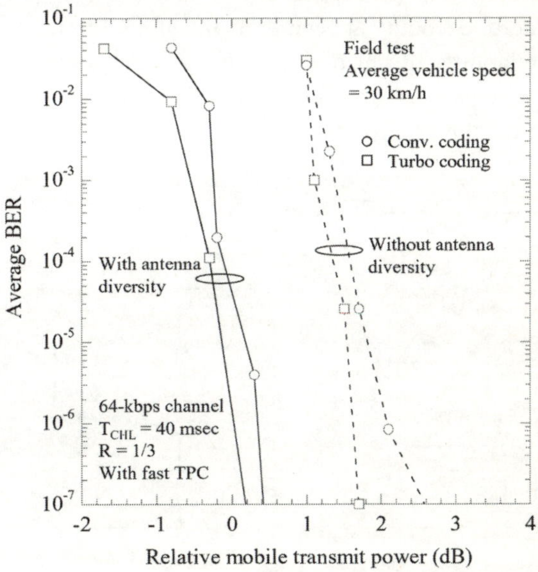

***Figure 13**. Average BER of 64-kbps data transmission with turbo coding as a function of mobile transmit power. Field experiments.*

i.e., superiority was confirmed to be above 1.0 dB. This abatement in the improvement with antenna diversity reception indicated that in an actual fading channel in the field experiments, the impact of path search for Rake combing and SIR measurement for fast TPC diminished the improvement in performance of the turbo coding due to a very low received signal power level.

VI.2. Site diversity (soft/softer handover)

Soft handoff or site diversity ("site diversity" hereafter) [23, 24], which was first implemented in the IS-95 CDMA standard [25], is an essential technique together with fast TPC in improving transmission impairment due to multipath fading and shadowing near the cell edge. The simplified configuration of site diversity is illustrated in Figure 14. In the forward link, the same original information sequences before channel coding are transferred to N BSs (N is the number of BSs with which the MS is associated) through the back haul (wired transmission line between BS and radio network controller (RNC)) from a RNC and transmitted from two BSs using different scrambling codes. The received signals after Rake combining at the MS are combined symbol-by-symbol with maximal ratio combining (MRC) followed by soft-decision Viterbi decoding. With inter-sector diversity in the reverse link, the Rake-combined signal of each sector is combined with MRC in the same way as in the forward link. On the other hand, with inter-cell diversity, in the reverse link, a hard-decision data sequence after soft-decision Viterbi decoding at each BS is transferred to the RNC via the back haul with the reliability information associated with each traffic channel. The transferred data sequences are selection-combined every selection period, according to the reliability information.

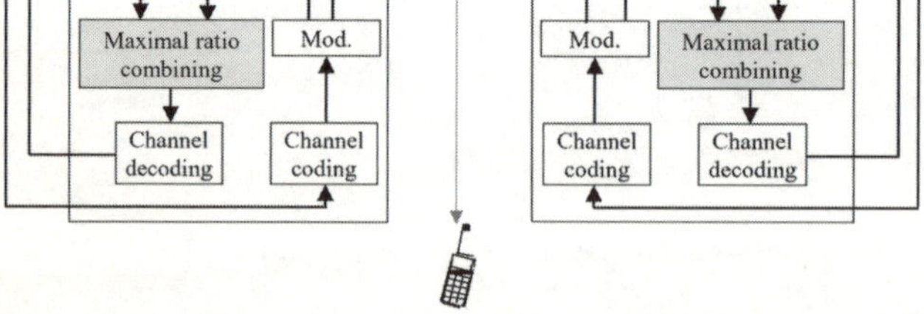

Figure 14. *Simplified configuration of site diversity.*

VI.2.1 Reverse link

The performance of reverse link inter-cell site diversity depends on the type of reliability information that is used. Therefore, we present a two-step SC scheme using two types of reliability information [26, 27]: CRC results calculated over selection interval T_{SEL} and the average received SIR measured over interleaving interval T_{ILV}. In our scheme, we use the number of slots with a measured SIR value greater than the target value of fast TPC, N_{SIR}, i.e., the number of TPC command bits to lower the transmit power during T_{ILV}, instead of the actual measured SIR value. This is because the transfer capacity in the back haul required for the reliability information of inter-cell site diversity can be significantly decreased (note that only 4 bits/frame are required for denoting the SIR average over one frame). The SC at the RNC was performed in two steps.

(1) Step 1: When multiple decoded data sequences transferred from N cell sites (BSs) indicate no CRC error, then the one data sequence over T_{SEL} among the data sequences yielding the successful CRC result is selected.

(2) Step 2: When all the CRC results transferred from N BSs indicate frame error, the data sequence during T_{SEL} with the larger N_{SIR} over T_{ILV} is selected.

The field experiments using inter-cell site diversity were conducted in an area near Tokyo in order to measure the BER performance in a 32-kbps data rate channel. The measurement course is a road running north and south, which passes through the middle of 2 BSs. The distance between BS 1 and BS 2 is approximately 2.5 km. The middle point of the measurement course is approximately 1,300 and 1,200 m apart from BS 1 and BS 2, respectively. On either side of the measurement course is a low-rise factory area. The view from BS 1 was line-of-sight (LOS) except at the end of the course, while it was non-line-of-sight (NLOS) from BS 2 due to the tall buildings. We set the soft-handover threshold to 3 dB. The difference in the measured average received signal powers from the two BSs was approximately 1 dB. Thus, the measurement course is a softhandover area within the prescribed threshold. The power delay profile with 1 (2) and 2 (1) paths were observed in the first half and the latter half of the course from BS 1 (BS 2). We set T_{SEL} = 10 msec. In the experiments, fast TPC was used only in the reverse link. The received signal power was set to be sufficiently high so that there was no TPC command bit error. The measured time variations of the instantaneous BER and received E_b/I_0 at BS 1 and BS 2, after inter-cell site diversity, and the mobile transmit power averaged over one radio frame length (=10 msec) are plotted in Figure 15. The target E_b/I_0 at each BS was set to 7 dB so that the average BER after intercell site diversity was approximately 10^{-3}. The figures show that bit errors occurred when the received E_b/I_0 at each BS dropped; however, the instantaneous received E_b/I_0 after inter-cell site diversity was maintained at almost a constant level. Therefore, the

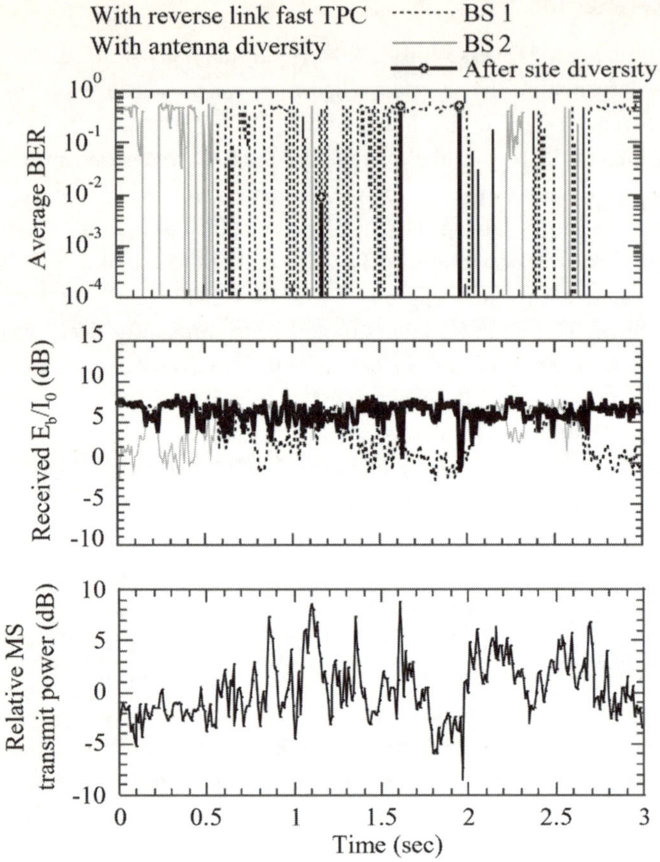

Figure 15.
Instantaneous time variations in the reverse link inter-cell site diversity. Field experiments.

measured BER after inter-cell site diversity significantly improved; nevertheless the BER measured at each BS was significantly degraded due to the reduced signal level caused by shadowing and fading variations. Since the target E_b/I_0 was set to satisfy the average BER of 10^{-3} after selection combining, burst error rarely occurred since convolutional coding was used.

VI.2.2 Forward link

When fast TPC is applied in the forward link inter-cell site diversity mode, each BS independently follows the TPC command bit sent from the MS via the reverse link. Therefore, the transmit power of each BS differs when a TPC command bit error occurs in the reverse link. An increase in the difference between the transmit powers of the BSs causes a reduction in site diversity gain and an increase in the interference to other users. To overcome this problem, several schemes that compensate for the BS transmit power were proposed [28, 29]. In the method proposed in [28], each BS controls its instantaneous transmit power by using a forgetting factor so that the difference between instantaneous transmit power and the BS-specific reference transmit power calculated by averaging the instantaneous values does not become large. However, it is difficult to quickly track variations in path loss including shadowing due to the movement of the MS. The method in [29] reduces TPC bit error by sending the same TPC bit over several slots in the site diversity mode, this prevents the transmit power difference between BSs from becoming large. However, in addition to the problem described in [29], the TPC delay increases. Therefore, we proposed the following two step algorithm to reduce the impact of TPC errors and keep the transmit power of the BSs the same as that shown in Figure 16 [26].

(1) First loop: the standard transmit powers, $P^{(k)}_{REF}$, of all BS_k are compensated by $\Delta P^{(k)}$ according to the dedicated control channel from a MS based on the average SIR measurement at a MS.

(16) $\Delta P^{(k)} = \text{Target\_}E_b/I_0 - \text{Measured\_total\_}E_b/I_0 \text{ (dB)}$,

where the Measured_total_E_b/I_0 and Target_ E_b/I_0 are the measured E_b/I_0 after Rake combining and the target E_b/I_0 at a MS, respectively. The $P^{(k)}_{REF}$ is constant during the length of G-slot and its value of n (= gxG)-th slot $P^{(k)}_{REF}(n)$ is updated every G-slot as $P^{(k)}_{REF}(g \times G) = P^{(k)}_{REF}((g-1) \times G) + \Delta P^{(k)}$.

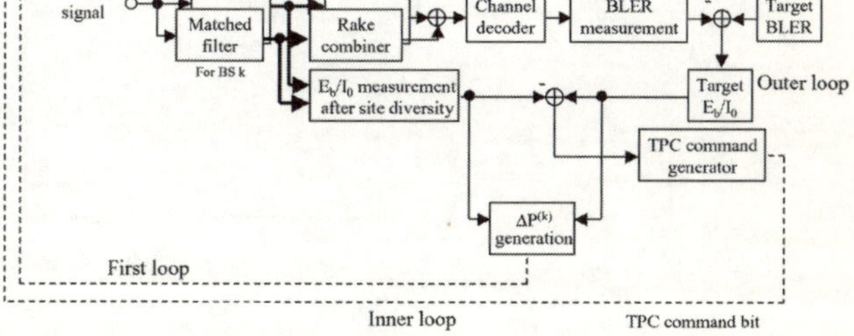

Figure 16. Combination of forward link site diversity and TPC.

(2) Second loop: the instantaneous transmit power, $P^{(k)}_{CL}(n)$, is controlled according to the TPC command bits (ΔTPC) by introducing the forgetting factor μ using the standard transmit power compensated in the first loop.

(17) $P^{(k)}_{CL}(n) = P^{(k)}_{REF} + \mu\,(P^{(k)}_{CL}(n-1) - P^{(k)}_{REF}) + \Delta_{TPC}$

The measured average BER performance in the forward link when inter-cell site diversity is applied is plotted in Figure 17 as a function of the total BS average transmit power. The measurement course and experimental conditions were the same as those in Figure 15. Fast TPC was used for the reverse and forward links. The number of maximum Rake fingers for the BS and MS was 4. The forgetting factor was set to R = 0.8. In the measurement course, it was observed that the instantaneous transmit power is controlled around the standard transmit power without dispersing to the maximum output during the course. The figure shows that the total transmit power of the 2 BSs at the average BER of 10^{-3} in the inter-cell site diversity is decreased by approximately 0.3 dB compared to a one cell-site connection. This improvement is small compared to that in the reverse link because the increase in interference due to transmissions from two BSs diminished the diversity effect. Thus, the site selection diversity transmit power control (SSDT) [30], in which only the primary BS transmits control bits to decrease the interference, was proposed.

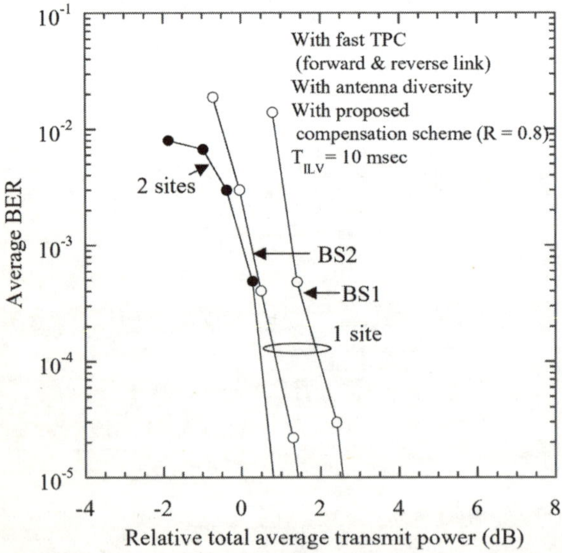

Figure 17. Average BER in the forward link inter-cell diversity. Fields experiments.

VI.3. Transmit diversity

Transmit diversity employing several antennas at a BS can improve the forward link transmission performance without increasing the complexity of the MS [31, 32]. Therefore, several transmit diversity schemes were adopted in the 3GPP standardization [5, 33]. The CPICHS with the same spreading code, but different data modulation patterns, are transmitted from two antennas in the same carrier phase. Two open-loop type transmit diversity schemes were adopted in the 3GPP standardization: TSTD [5] and space-time transmit diversity (STTD) [5, 34]. STTD, which is used for the CCPCH, transmits two data sequences in parallel after coding from two antennas using different channelization codes. Since the fading correlation between the two antennas is low, the fluctuation in the received signal level due to fading is mitigated. The operational principle and coding scheme of STTD is illustrated in Figure 18(a) and Figure 18(b), respectively. Let $S(m)$ be the QPSK symbol data sequence denoted as $S(m) = \exp j\phi(m)$, where $\phi(m) \in \{h\pi/2 + \pi/4\ ;\ h = 0 - 3\}$ is the QPSK modulation phase. Then, two successive symbols, $S(m)$ and $S(m + 1)$ are treated as a pair, where m denotes an even number. The two symbol sequences, $d^1(m)$ and $d^2(m)$, for antennas #1 and #2 generated in the STTD encoder are expressed respectively as

(18a) $d^1(m) = S(m)$ and $d^1(m + 1) = S(m + 1)$,

(18b) $d^2(m) = -S^*(m + 1)$ and $d^2(m + 1) = S^*(m)$

It is clear that the orthogonality between the two data sequences is maintained irrespective of the spreading code sequence.

(a)

(b)

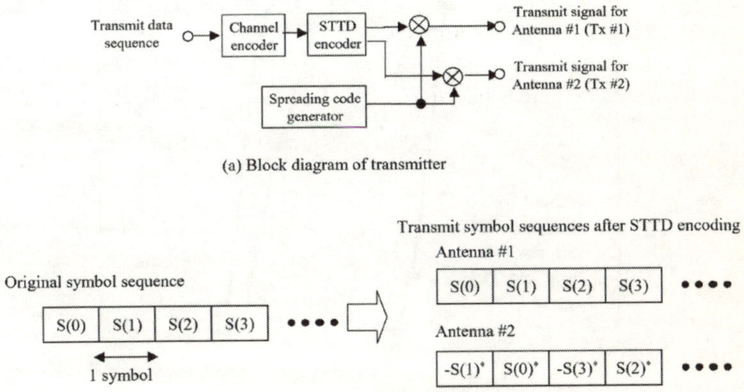

***Figure 18**. Operational principe of STTD.*
(a) Block diagram of transmitter, (b) example of STTD encoding.

On the other hand, closed-loop type (two modes were standardized) transmit diversity is used for DPCHs, in which the transmit antenna weights are controlled by the FBI generated at the MS [33]. Let $W_1 = A_1e^{j\phi 1}$ and $W_2 = A_2e^{j\phi 2}$ be the transmit antenna weights. Thus, in Mode 1, the transmitted phase of the second antenna, ϕ_2, is changed with the accuracy of $\pi/4$ according to the FBI from the MS so that the received SIR after combining is maximum. This is expressed as $\phi_1 = 0$, $\phi_2 = \{+/- \pi/4, +/- 3\pi/4\}$, $A_1 = A_2 = \sqrt{1/2}$. Meanwhile, the transmitted amplitudes of two data sequences are also controlled by FBI bits as well as the transmitted carrier phase in Mode 2.

The measured average BER performance with STTD is plotted in Figure 19 when fast TPC was not applied in the forward link as a function of the average received E_b/I_0, where I_0 is the multipath interference plus background noise power density [35]. The measurement course was course #1 described in [19]. The performance with and without antenna diversity reception at a MS is shown in the figures. The BER performance with signal antenna transmission is also depicted for comparison. Figure 19 shows that the average required received E_b/I_0 at the average BER 10^{-3} with STTD was decreased by approximately 1.5 (1.0) dB without (with) antenna diversity reception. The improvement using STTD with antenna diversity reception became smaller than that without antenna diversity because the degradation of the channel estimation due to a lower

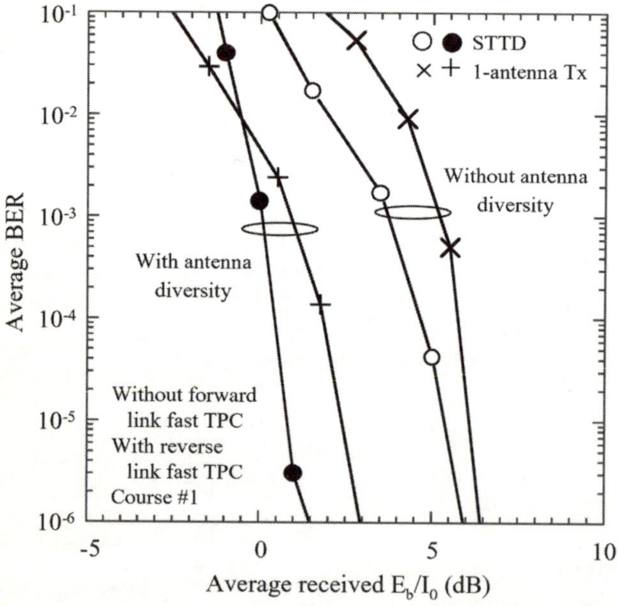

Figure 19. Average BER performance in field experiment (without forward link fast TPC).

received level offset the additional diversity effect by STTD when using Rake path diversity and antenna diversity reception. From the figure, the effectiveness of STTD for a channel without TPC such as a common control channel was elucidated in a real multipath-fading channel.

VII. W-CDMA capacity enhanced technologies

In DS-CDMA systems, due to multipath fading and shadowing as well as distance-dependent path loss, severe MAI is often produced, which significantly reduces the link capacity. In the forward link, although the orthogonality among the same propagation channels is achieved by using OVSF channelization codes, the MPI especially from high rates users is severe. Thus, the IC or multiuser detection (MUD) [36, 37] and adaptive antenna array receiver [38, 39] are promising techniques to reduce MAI and thereby decrease the transmit power of MSs in the reverse link. AAA-TD [40-43] is effective in decreasing severe MPI in the forward link without changing the air interface and adding complexity to the MS. High-capacity BS transceiver configuration examples are shown in Figure 20. Figure 20(a) shows the configuration using CAAAD [44, 45] and AAA-TD in the reverse and forward links, respectively. Figure 20(b) is a configuration including IC and AAAD. In the reverse link, the COMSIC [46-48] and CAAAD both with multiple antenna configurations of greater than two are installed in parallel. Only the COMSIC with multiple antenna reception is used for data demodulation in the reverse link, while the CAAAD receiver is used only to generate the transmit antenna weights for AAA-TD, which are created by performing both RF circuitry calibration and carrier frequency calibration on the receiver antenna weights. However, by using the parallel structure of COMSIC and CAAAD, two advantages are obtained. The first is, since the interference replicas of other users for each array antenna are independently generated in COMSIC, high-quality decoded data symbols at the COMSIC output are used to update the transmitter (receiver) weights in addition to pilot symbols. The second is that SIR measurement for fast TPC associated with outer-loop control and accurate path search is possible at the CAAAD output after MAI and MPI suppression.

VII.1. COMSIC

VII.1.1. COMSIC receiver

The COMSIC is categorized into two types: the serial (successive)-type and the parallel-type. The parallel-type COMSIC is more practical than the serial-type due to its inherent advantage of a short processing delay, which is derived from the

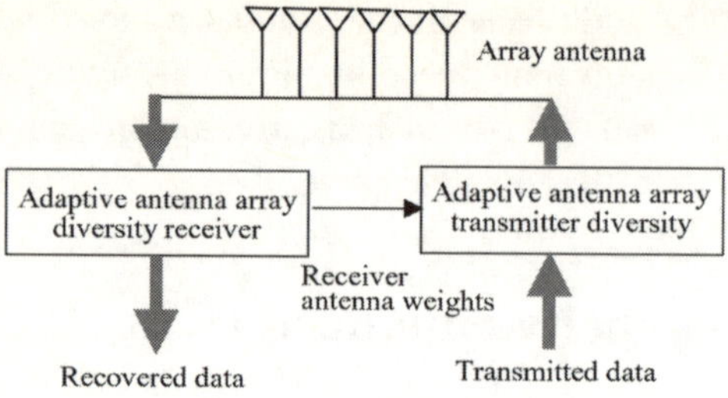

(a) RL: CAAAD – FL: AAA-TD

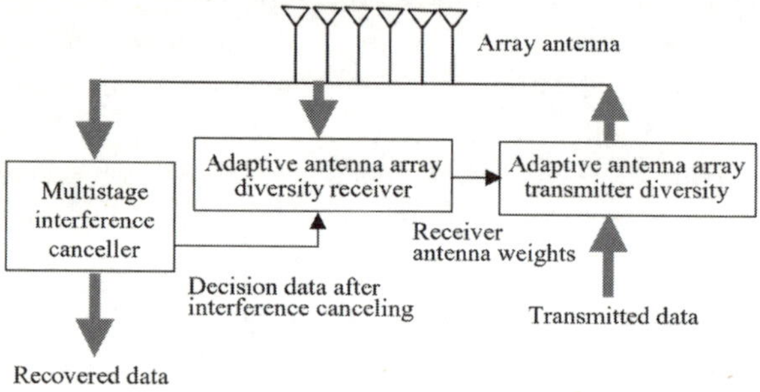

(b) RL: IC – FL: AAA-TD

Figure 20. *High-capacity BS transceiver configuration examples.*

despreading, PSA-channel estimation, Rake combing, and generation of interference replicas for all active users simultaneously performed in parallel. The serial-type COMSIC, on the other hand, is accompanied by a long processing delay according to the number of active users because the subtraction of generated interference replicas is successively conducted in the order of the received signal power of each user. The interference suppression effect of the parallel-type COMSIC is inferior to that of the serial-type due to its poor channel estimation accuracy and data detection error because channel estimation and data decision must be performed under severely low SIR conditions at the conventional MF-based Rake receiver in the first stage. In order to improve the interference suppression effect to the same level as that of the serial-type COMSIC, the channel estimation accuracy must be improved and the data decision error must be decreased to generate accurate interference replicas. Therefore, we proposed an improved parallel-type COMSIC with iterative channel estimation (ICE) using both pilot and decision-feedback data symbols in order to improve the interference suppression effect to the same level as the serial-type COMSIC [48]. Our scheme is based on recursive channel estimation at each canceling stage to improve the accuracy of the generated interference replicas and the iterative channel estimation is applied at each stage. The recursive channel estimation associated with iterative channel estimation at each stage significantly improves the interference suppression effect of the parallel-type COMSIC almost to the same level as that of the serial-type.

Figure 21 is a block diagram of the parallel-type COMSIC. Composite signal sample sequences received at two antennas are despread using MFs. At each stage, there are K channel estimation and interference replica generation units (CEIGUs), where K is the number of active users. Let $I_{k,b,l}$ be the MAI received via the l-th path ($1 \leq l \leq L_k$) of the b-th antenna ($1 \leq b \leq 2$) of the k-th user ($1 \leq k \leq K$)in the p-th stage ($1 \leq p \leq P$) and $\hat{I}^{(p)}_{k,b,l}$ be its estimated value. In the first stage, the received signal of each antenna is directly embedded into the CEIGU as well as the conventional MF-based Rake receiver. For the CEIGU input signal of the b-th antenna of the k-th user in the p-th stage, ($p \geq 2$), the estimated interference replicas of all other users, $\sum_{\substack{i=1 \\ i \neq k}}^{K} \sum_{j=1}^{L_i} \hat{I}^{(p-1)}_{i,b,j}$, and other multipaths of the user's own channel, $\sum_{\substack{j=1 \\ j \neq l}}^{L_k} \hat{I}^{(p-1)}_{k,b,j}$, which are estimated in the previous stage, are weighted by the interference rejection weight control (IRWC) value, α_p [47], and removed from the signal received. It was confirmed by the some of the authors [47] that by setting α_p to less than one, the BER performance using COMSIC is improved since the influence of the generation error of the interference replica due to the channel estimation error and data decision error is mitigated. The effect of IRWC is larger in the parallel-type COMSIC than in the serial-type. In our COMSIC scheme, since the complex fading envelope (channel impulse response) is recursively estimated and updated at each cancelling stage, the accuracy of channel estimation and thereby that of the regenerated interference replica is successively improved.

Figure 22 is a block diagram of the proposed CEIGU. In the proposed scheme, the ICE using both pilot and decision-feedback data symbols after channel decoding (hereafter forward error correction (FEC) decoding) [48] is applied to each stage. Each CEIGU performs channel estimation, Rake combining, de-interleaving, Viterbi decoding, data decision, and replica generation of the MAI for each user. Let $\xi_{k,b,l}(t)$, be the complex-valued channel impulse response of the l-th path of the b-th antenna of the k-th user signal and $\hat{\xi}^{(p,r)}_{k,b,l}(n)$ be its estimated value after the r-th iteration ($0 \leq r \leq R$) in the p-th stage. In the first stage, the initial channel estimate (estimated complex channel impulse response) $\xi^{(1,0)}_{k,b,l}(n)$ of the n-th slot is computed by coherently averaging $2N_p$ pilot symbols included in two successive slots. Then, a Rake combined data sequence is obtained by coherently summing the $L_k \times B$ resolved signal components using

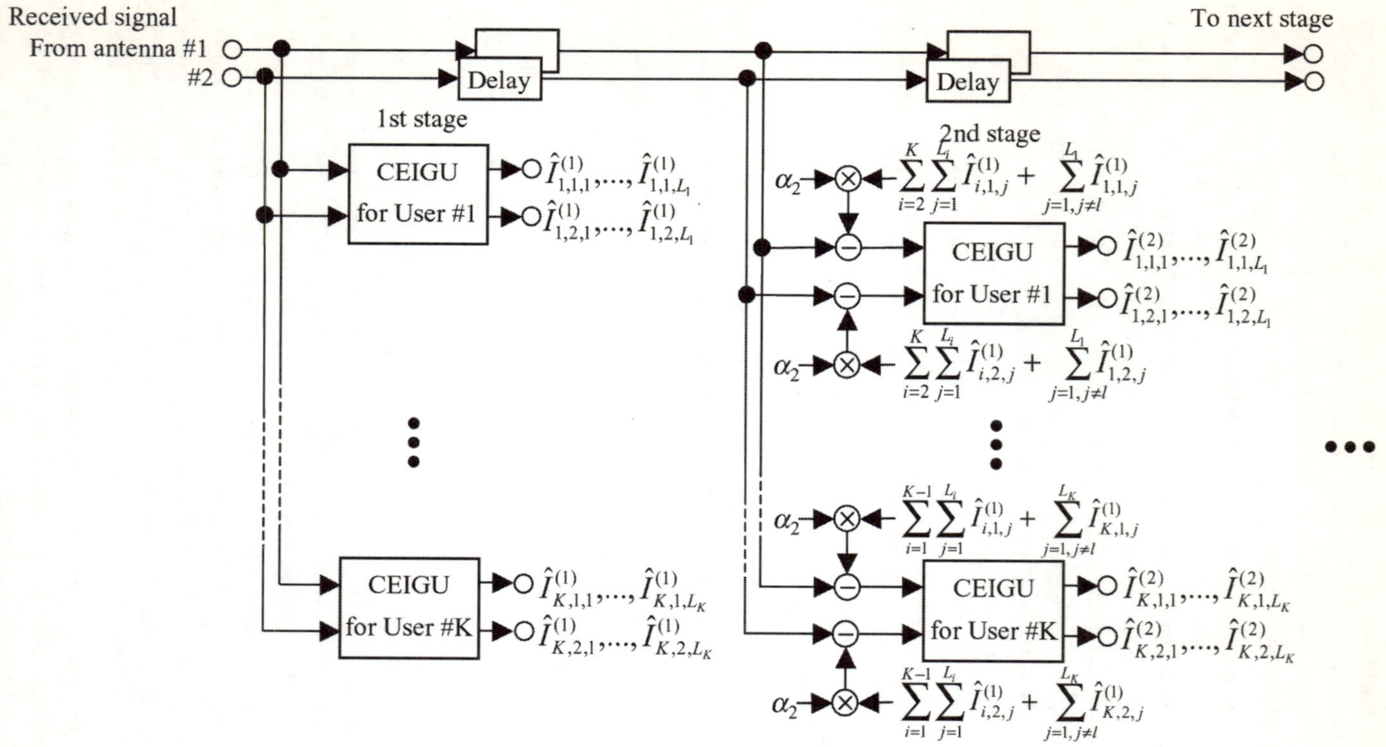

Figure 21. *Block diagram of parallel-type coherent multistage interference canceller.*

channel estimates $\hat{\xi}_{k,b,l}^{(1,0)}(n)$. In ICE using both pilot and decision-feedback data symbols after FEC decoding (hereafter ICE after FEC decoding), the Rake combined sequence, $\hat{d}_k^{(1,1)}(n, m)$, is de-interleaved and soft-decision Viterbi decoded to recover the transmitted binary data sequence, $\hat{b}_k^{(1,1)}(i)$. The original data sequence recovered is convolutionally-encoded and interleaved to generate the binary decision-feedback data symbol sequence, $\tilde{d}_k^{(1,1)}(n, m)$. Then, after the data modulation of the MF output signal is removed by reverse-modulation using the decision-feedback data symbols, channel estimation using N_d decision-feedback data symbols in addition to $2N_p$ pilot symbols is performed for each slot. On the other hand, in ICE using both pilot and tentative-decision data symbols before FEC decoding (hereafter ICE before FEC decoding), we use the binary tentative-decision data sequence after Rake combining (namely before FEC decoding), $sgn[\hat{d}_k^{(1,1)}(n, m)]$, instead of the decision-feedback data sequence after FEC decoding in ICE after FEC decoding. Let $\hat{\xi}_{k,b,l}^{(1,R)}(n)$ and $\tilde{d}_k^{(1,R)}(n, m)$ be the channel estimate after the *r-th* iteration ($1 \leq r \leq R$), then the interference replica of each path of each antenna is re-generated precisely as

$$(19) \quad \hat{I}_{k,b,l}^{(1)}(t-\hat{\tau}_{k,b,l}) = \hat{\xi}_{k,b,l}^{(1,R)}(t-\hat{\tau}_{k,b,l}) \cdot \tilde{d}_k^{(1,R)}(t-\hat{\tau}_{k,b,l}) \cdot c_k(t-\hat{\tau}_{k,b,l}).$$

By using the generated interference replicas, the MF input of the CEIGU of the *l-th* path of the *b-th* antenna of the *k-th* user in the second stage is represented as

$$(20) \quad r_{k,b,l}^{(2)}(t) = r_b(t) - \alpha_2 \cdot \left[\sum_{\substack{i=1 \\ i \neq k}}^{K} \sum_{j=1}^{L_i} \hat{I}_{i,b,j}^{(1)}(t-\hat{\tau}_{i,b,j}) + \sum_{\substack{j=1 \\ j \neq l}}^{L_k} \hat{I}_{k,b,j}^{(1)}(t-\hat{\tau}_{k,b,j}) \right].$$

Furthermore, by using the interference replica generated in the previous stage, the MF input of the CEIGU in the *p-th* stage ($p \geq 2$) is anticipated as

$$(21) \quad r_{k,b,l}^{(p)}(t) = r_b(t) - \alpha_p \cdot \left[\sum_{\substack{i=1 \\ i \neq k}}^{K} \sum_{j=1}^{L_i} \hat{I}_{i,b,j}^{(p-1)}(t-\hat{\tau}_{i,b,j}) + \sum_{\substack{j=1 \\ j \neq l}}^{L_k} \hat{I}_{k,b,j}^{(p-1)}(t-\hat{\tau}_{k,b,j}) \right],$$

where

$$(22) \quad \hat{I}_{i,b,j}^{(p)}(t-\hat{\tau}_{i,b,j}) = \hat{\xi}_{i,b,j}^{(p,R)}(t-\hat{\tau}_{i,b,j}) \cdot \tilde{d}_i^{(p,R)}(t-\hat{\tau}_{i,b,j}) \cdot c_i(t-\hat{\tau}_{i,b,j}).$$

After R iterations, data sequence $\hat{d}_k^{(P,R+1)}(n, m)$ is finally de-interleaved and soft-decision Viterbi decoded to recover the refined decoded binary data sequence, $\hat{b}_k^{(P,R+1)}(i)$.

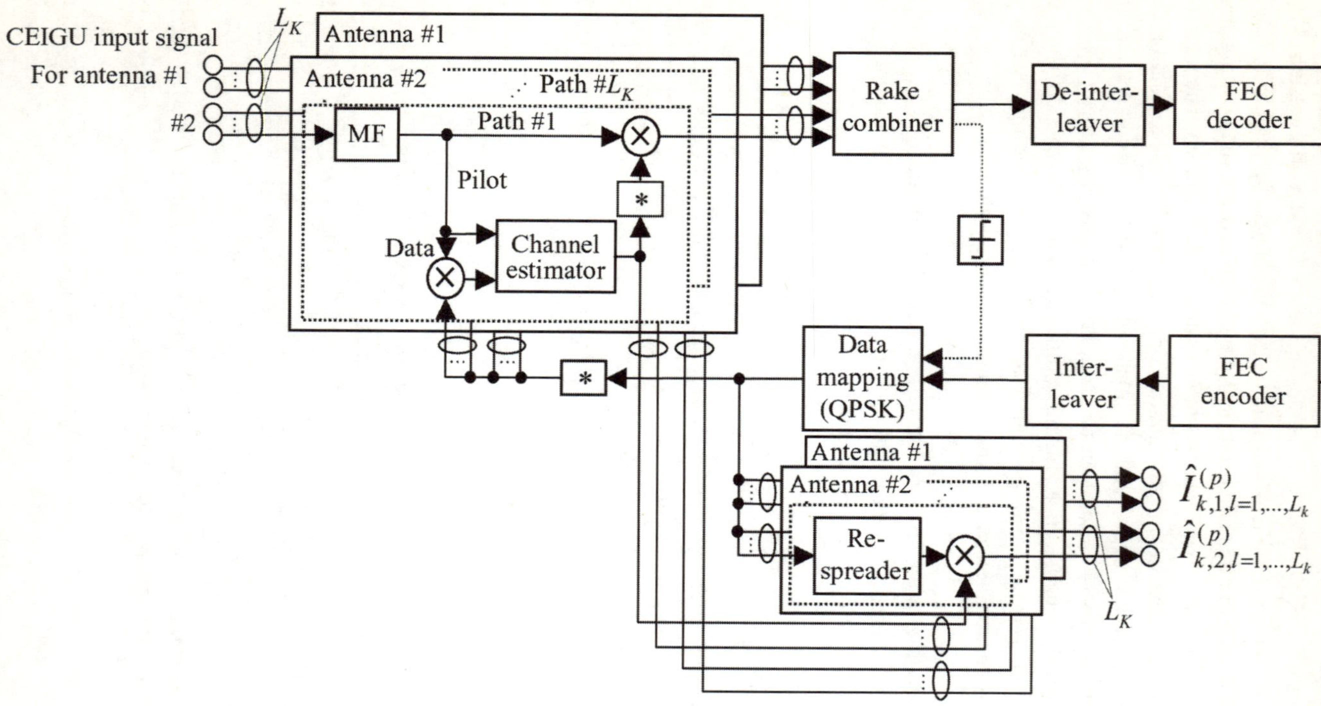

Figure 22. *Block diagram of* CEIGU *with* ICE *using pilot and decision-feedback data symbols.*

VII.1.2. Experimental results

The required average transmit E_b/N_0 for achieving the average BER of 10^{-3} of the parallel-type COMSIC using two-branch antenna diversity reception is plotted in Figure 23 as a function of f_D based on laboratory experiments using a fading simulator. External Gaussian noise was added as background noise to the receiver input. The number of active users was K = 3, and the SF of the desired user and two-interfering users were SF = 64 and 8, respectively. The number of stages was P = 3. The power delay profile used here was a two-path profile, each path being subjected to independent Rayleigh fading with equal average power. The target E_b/I_0 ratio of the desired user to interfering user was set $\Delta E_b/I_0 = -3$ dB and – 6 dB. Thus, when $\Delta E_b/I_0 = -6$ dB, the received signal power of the interfering user is 32-fold that of the desired user. The TPC delay was assumed to be one slot. Figure 23 shows that the transmit E_b/N_0 required using decision data symbols after Rake combining with P = 3 is slightly decreased compared to that using decision data symbols after channel decoding with P = 2 irrespective of f_D. The transmit E_b/N_0 required using decision data symbols after Rake combining with P = 2 is reduced by approximately 0.6 (0.8) and above 2.0 dB compared to that of COMSIC using only pilot symbols and the MF-based Rake receiver, respectively, when $\Delta E_b/N_0 = -3$ (– 6) dB. Furthermore, the loss of the required E_b/N_0 from the single user case due to the channel estimation error and

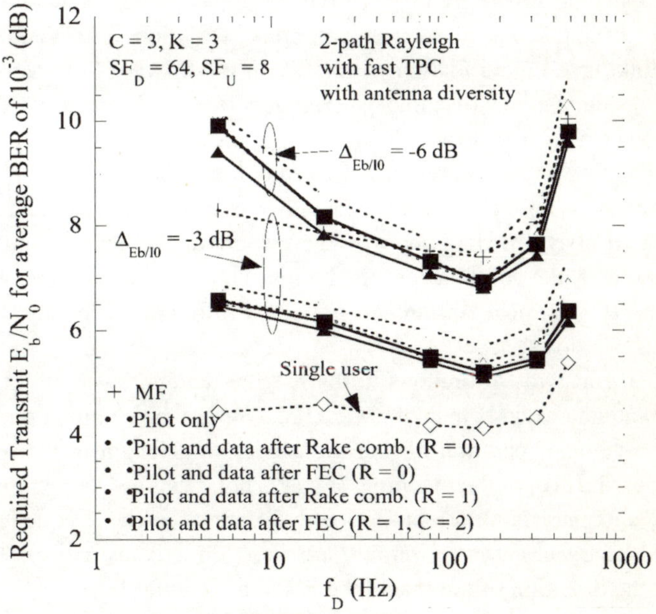

Figure 23.

Average transmit E_b/N_0 for the average BER 10^{-3} using parallel type comsic as a function of f_D (with antenna diversity reception, $\Delta E_b/I_0 = -3, -6$ dB).

data decision error when $\Delta E_b/N_0 = -3$ dB was approximately 0.7 to 2.0 dB. When the f_D was small, the transmit E_b/N_0 was slightly increased so that the fast TPC should compensate for the drop in the signal level received due to fading. Then, although the required transmit E_b/N_0 was decreased as f_D was increased owing to the increasing interleaving effect associated with channel coding, it was significantly degraded due to the degradation in the channel estimation tracking ability against fast fading when f_D was larger than approximately 150 Hz.

VII.2. AAAD

VII.2.1. CAAAD receiver/AAA-TD

We also investigated the interference suppression effect in the reverse and forward links using the experimental CAAAD receiver/AAA-TD [40-45]. The carrier frequencies were the same as those of the COMSIC. The spreading chip rate was 4.096 Mcps and the bandwidth was 5 MHz. Figure 24 shows an overall block diagram of the four-antenna AAAD transceiver. The receiver antenna weights in the reverse link are generated in the CAAAD receiver block. In our scheme, transmitter antenna weights are generated by modifying the receiver antenna weights (i.e., performing calibrations we will be explaining later) [40-43]. Our concern is frequency division duplex (FDD) in W-CDMA systems, thus, the instantaneous phase and amplitude variations due to fading in the reverse link have no correlation with those in the forward link. Noting that the distance-dependent macroscopic propagation factors determining the average signal power, i.e., path loss and shadowing, are not frequency dependent, we designed the CAAAD receiver based on the following concept. The adaptive antenna array forms an antenna beam that tracks only slow changes in the directions of arrival (DOAs) and average powers of the desired and interfering users, and the Rake combiner tracks the instantaneous variations in the channel conditions to maximize instantaneous signal-to-interference plus background noise power ratio (SINR).

We perform two calibrations to the receiver antenna weights in generating transmit antenna weights in the forward link. We use the common array antennas with the antenna separation of half the carrier wavelength in the reverse link. Thus, when the DOA of the incoming signal is not 0 degrees, the phase difference between antennas is different between different carrier frequencies. If the generated receiver antenna weights are used directly as transmitter antenna weights, the direction of the main lobe in the transmitter beam pattern is shifted from the original DOA of the desired user and the directions of the beam nulls are also shifted from those of interfering users. Therefore, in our approach, we shift the direction of the main lobe in the transmitter beam pattern so that it coincides

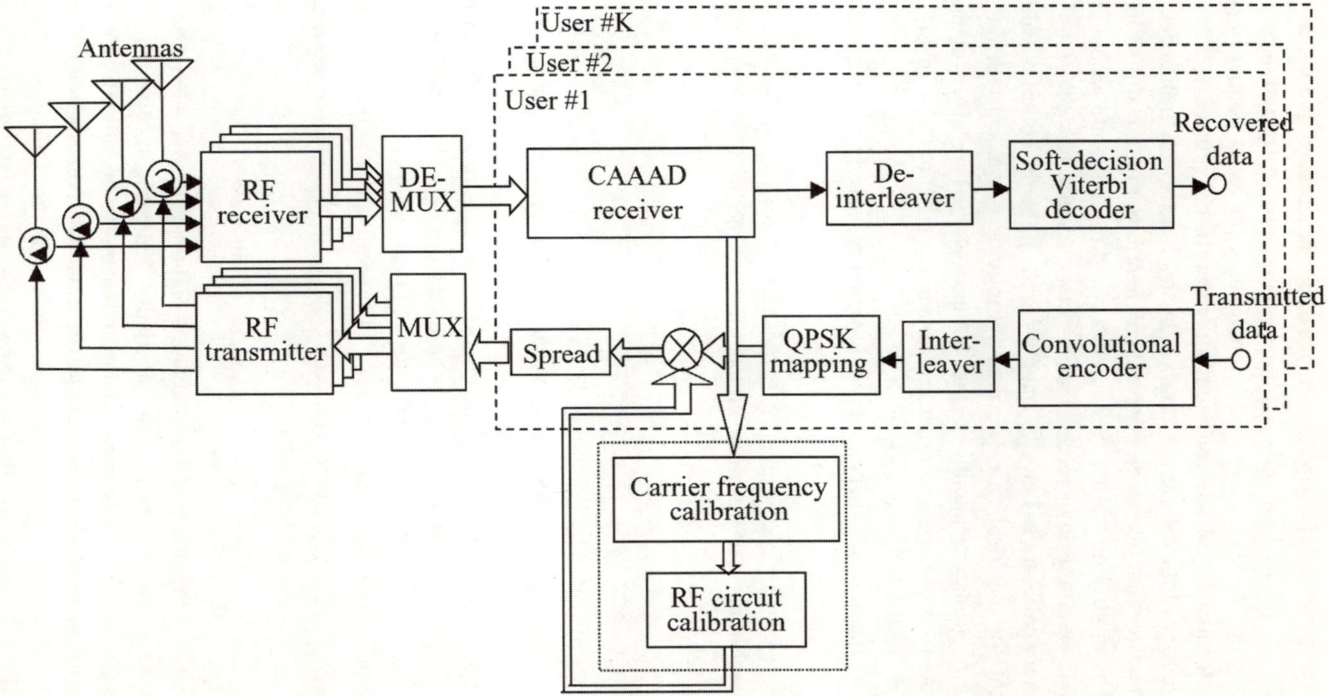

Figure 24. *Overall block diagram of* CAAAD *transceiver.*

with the main lobe in the receiver beam pattern. We call this modification in the transmitter beam pattern carrier frequency calibration.

On the other hand, since adaptive antenna array processing is done in the baseband, the generated reverse link weights are reflected on the phase/amplitude deviations in the transfer functions of the radio frequency (RF) receiver circuitry associated with different antennas. In the transmitter, the RF transmitter circuitry of the different antenna branches is adversely influenced by the different transfer functions (see Figure 24). Therefore, the transmit weights are generated by modifying the receiver antenna weights taking into account the phase/amplitude variations in RF receiver/transmitter circuitry. This calibration is called RF circuitry calibration. Let $w^{(i)}_{ideal}$, and $w^{(i)}_{R}$ and $x^{(i)}_{RX}$ be the complex-valued antenna weight for the ideal case (no phase/amplitude errors exist in the RF receive circuitry transfer functions), the complex-valued receiver antenna weight generated in CAAAD, and complex-valued transfer function of the RF receive circuitry of the *i-th* antenna branch, respectively. For the ideal case (i.e., $x^{(i)}_{RX}$ = const. for all antenna branches), received signal $r^{(i)}_{RX}$ should be weighted by $w^{(i)}_{ideal}$ to be combined for beam forming. However, in a real receiver, $r^{(i)}_{RX}$ goes through the RF receiver circuitry and then, weighted by $w^{(i)}_{R}$, produces $w^{(i)}_{R}\ x^{(i)}_{RX}\ r^{(i)}_{RX}$, before combining. The resultant signal must be the same as that of the ideal case. Therefore, we obtain

(23) $$w^{(i)}_{ideal} = w^{(i)}_{R} x^{(i)}_{RX}$$

The weights provided by Equation (23) cannot be directly used to form the transmit beam because the signal weighted at the baseband stage suffers phase/amplitude shift due to the RF transmit circuitry before transmission from an antenna. Let $x^{(i)}_{TX}$ be the transfer function of the RF transmit circuitry for the *i-th* antenna branch. The signal to be transmitted from the *i-th* antenna is first weighted using transmit antenna weight $w^{(i)}_{T}$ and then suffers phase/amplitude shift (equivalent by multiplication of $x^{(i)}_{TX}$). The equivalent antenna weight becomes, therefore, $w^{(i)}_{T} \cdot x^{(i)}_{TX}$, and this must be equal to $w^{(i)}_{ideal}$. As a consequence, we obtain

(24) $$w^{(i)}_{T} = w^{(i)}_{ideal} / x^{(i)}_{TX} = w^{(i)}_{R} \left(x^{(i)}_{RX} / x^{(i)}_{TX}\right)$$

Using Equation (24), the transmit antenna weights (at the base band beam forming stage) can be obtained from adaptively generated receive antenna weights. Both $x^{(i)}_{RX}$ and $x^{(i)}_{TX}$ can be measured even during the operation mode. By performing the above two calibrations on the generated receiver antenna weights in the CAAAD receiver, the maximum gain is obtained toward the desired signal direction and the nulls are nearly directed towards the interfering sources.

Figure 25 shows a block diagram of the digital beam former and Rake combiner in the CAAAD receiver. This block comprises MFs, a beam former, PSA-coherent Rake combiner, and weight controller. The MF output signal samples of all antennas for each resolved path are weighted by the receiver antenna weight, which is common to all paths for Rake combining, and then combined. Since it was reported that the angle spread among paths is within ten degrees in an urban area with a high elevation antenna configuration such as in a cellular system [49, 50], we used the common antenna weights, i.e., common receiver beam pattern, for all paths. The resulting composite channel gain at the beam former output is estimated by using the pilot symbols multiplexed into the Q-channel of three successive slots for coherent Rake combining of different resolved paths. The receiver antenna weights in the CAAAD receiver are updated so that the mean squared error (MSE) between Rake-combined signal and the reference signal is minimized. As a reference signal, we used the decision-feedback data symbols after FEC decoding in addition to pilot symbols for generating refined MSE and improving channel estimation accuracy.

VII.2.2. Experiments

First, we investigated the effect of receiver antenna weight updating using decision-feedback data symbols after FEC decoding in addition to pilot symbols and performance difference for channel coding. The average BER performance measured with the CAAAD receiver with fast TPC using convolutional coding and turbo coding is shown in Figure 26 as a function of the average transmit E_b/N_0 in the reverse link. The DOAs of the desired and interfering users were set to 0 and + 40 degrees, respectively, in a 2-user environment. The ratio of the target E_b/I_0 for fast TPC of the desired user to interfering user was $\Delta E_b/I_0 = -12$ dB. The performance with receiver weight updating using decision-feedback data symbols as well as pilot symbols, and pilot symbols only are plotted as solid and dotted lines, respectively. It was assumed that f_D is 5 Hz and T_{CHL} is 20 msec. Figure 26 shows that the transmit E_b/N_0 required for satisfying the average BER of 10^{-3} and 10^{-6} with the CAAAD receiver using decision feedback antenna weight updating and channel estimation is decreased by approximately 0.8 dB compared with the case using only pilot symbols both for convolutional coding and turbo coding. These results also confirm that for T_{CHL}= 20 msec, the required transmit E_b/N_0 at the average BER of 10^{-6} of the CAAAD receiver using decision feedback antenna weight-updating and channel estimation with turbo coding is decreased by approximately 0.5 dB compared to the case with convolutional coding. Thus, the results given here agree well with those from the MF-based Rake receiver in Figure 13.

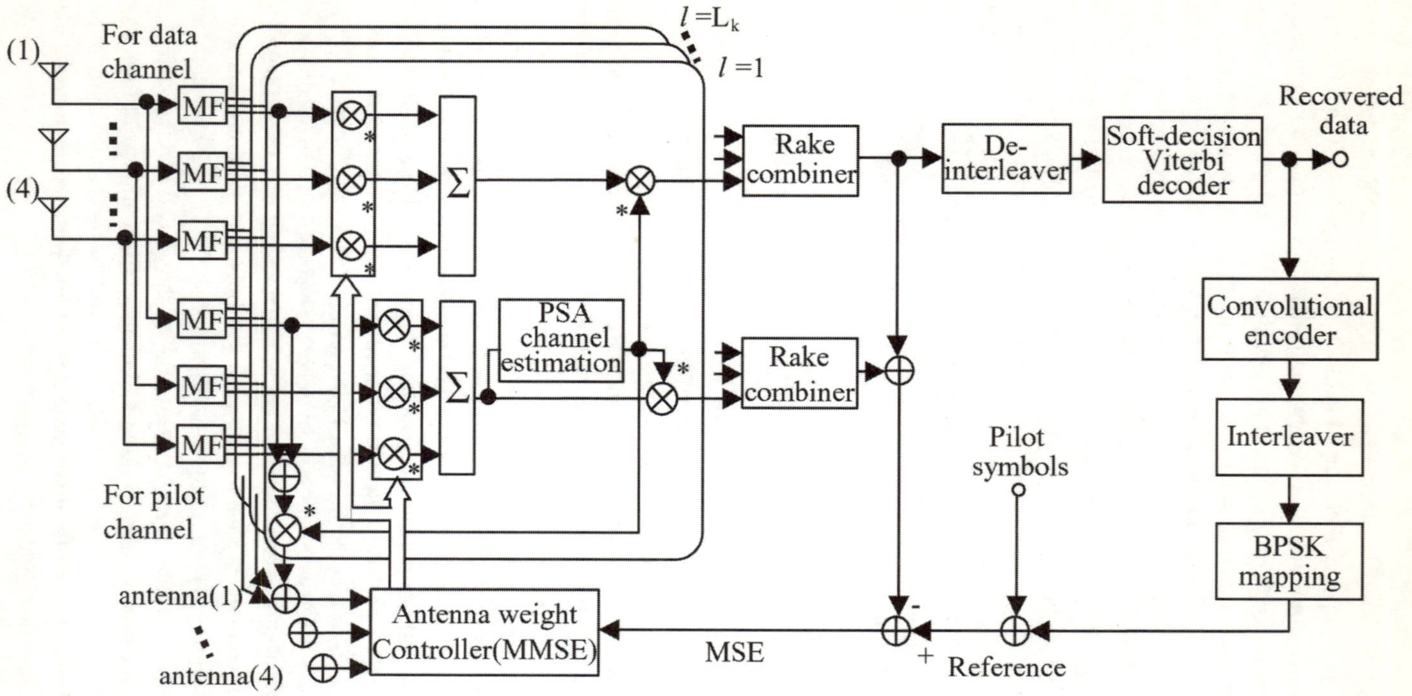

Figure 25. *Block diagram of CAAAD processing.*

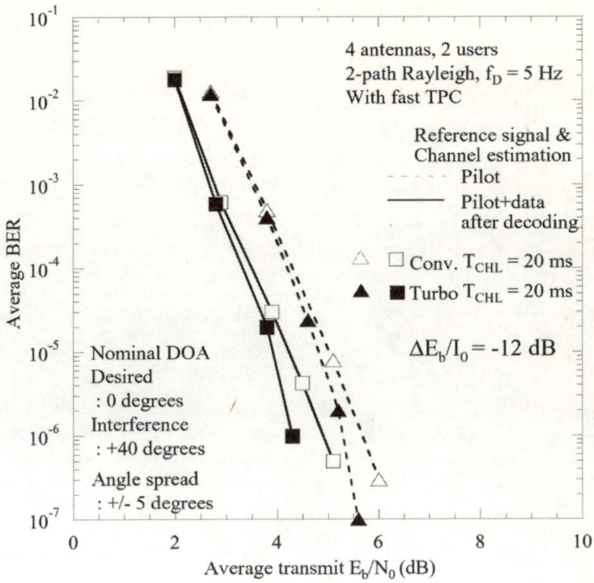

Figure 26.
Average BER performance as a function of average transmit E_b/N_0 (with fast TPC, 2 users).

Moreover, we evaluated, based on laboratory experiments, the forward link performance using four-antenna AAA-TD with SIR based fast TPC. The average BER performance of a MS using AAA-TD when fast TPC was used in the forward link is plotted in Figure 27 as a function of the average transmit E_b/N_0. The experimental configuration of AAA-TD is described in [43]. The number of users was four (3-interfering users) and the DOAs of the desired and interfering users were set to $\theta_D = -50$ degrees and $\theta_U = -20, +15, +45$ degrees. The BER performance for AAA-TD (when carrier frequency calibration was applied) is shown and the performance with one antenna transmitter is also plotted for comparison. The ratio of the target E_b/I_0 of the desired user to the interfering users, $\Delta E_b/I_0$, is – 3, – 5 and – 12 dB and the single-user case is also shown. Figure 27 clearly shows that when using the 1-antenna transmitter, as the interfering power is increased, the BER performance is significantly degraded, i.e., error floor is observed, due to the severe MPI from the interfering users. Meanwhile, by applying AAA-TD, almost identical BER performance was achieved even for the case of $\Delta E_b/I_0 = -12$ dB owing to the significant interference suppression effect. This result demonstrates that the RF circuitry calibration and carrier frequency calibration are working satisfactorily and that consequently AAA-TD is very effective in decreasing strong MPI especially from high-rate channels.

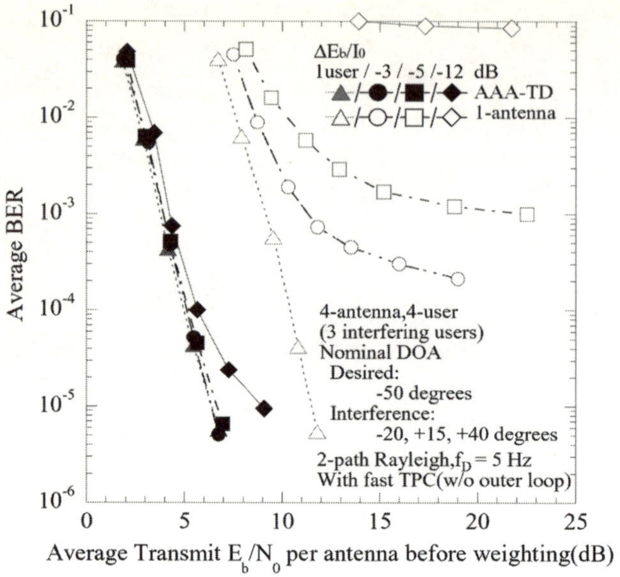

Figure 27. *Average* BER *performance of* AAA-TD *as a function of average transmit* E_b/N_0 *(with fast* TPC*).*

Next, the results of field experimental conducted near Tokyo are evaluated. In the CAAAD receiver experiments, MS 1 (desired user) moved along the measurement course, which was located approximately 600 to 850 m away from the BS. The height of the BS antenna was 50 m from the ground and 120-degree sectored antennas were used. When MS 1 moved along the measurement course, the DOA toward the BS changed from –10 to +10 degrees. In the measurement course, one-to-two path fading appeared in the first half of the course followed by two-to-three path fading with an average power difference of approximately 3 dB. On the other hand, MS 2 (interfering user) was located at a fixed point 600 m away from the BS and almost in the LOS path (thus, single-path). The DOA of MS 2 was approximately + 40 degrees. The measured BER performance is plotted in Figure 28 as a function of the average received signal power with the average received SIR of MS 1 as a parameter. For comparison, the results of space diversity using MRC are also plotted (antenna separation was 10λ). Figure 28 shows that the required average received signal power for obtaining the average BER of 10^{-3} is decreased by approximately 8 to 10 dB using the CAAAD receiver compared to using space diversity.

The average BER performance measured at a MS using AAA-TD in the forward link is plotted in Figure 29 as a function of the average received signal power

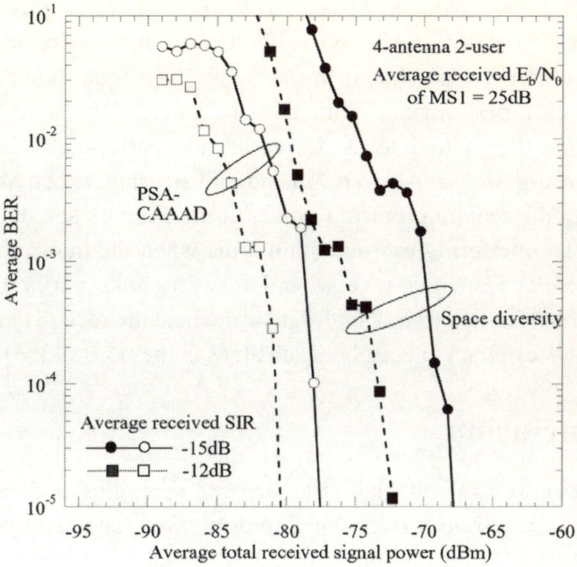

***Figure 28**. Average BER performance as a function of average received signal power. Field experiment.*

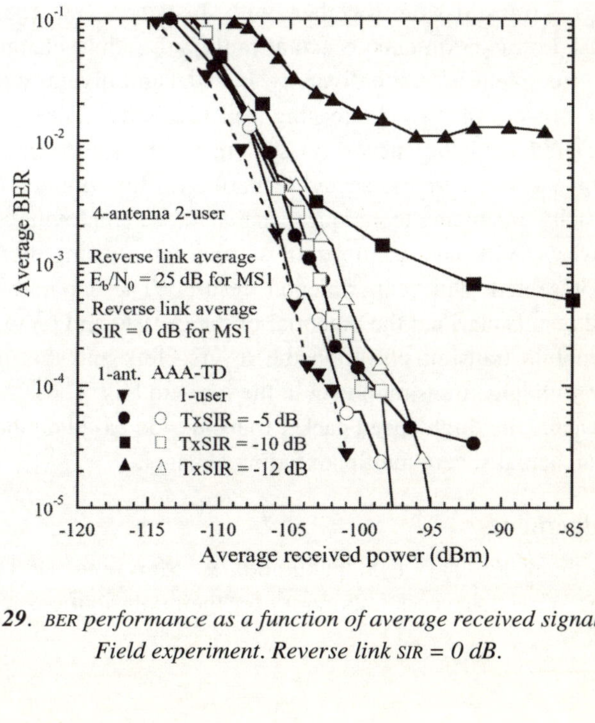

***Figure 29**. BER performance as a function of average received signal power. Field experiment. Reverse link SIR = 0 dB.*

when the average received SIR of MS 1 is 0 dB. The transmitted SIRs of MS 1 before weighting are – 5, – 10, and – 12 dB. The BER performance of a one-antenna transmitter is also shown in the figure. The figure clearly shows that although the BER performance with the one-antenna transmitter is severely degraded as the transmitted SIR is decreased, the performance is significantly improved by using AAA-TD due to beam and null steering. When AAA-TD is used, the increase in the required transmit power at the average BER of 10^{-3} from the case without an interfering user is within 5 dB when the transmitted SIR is –12 dB. These results from the reverse and forward links verify the effects of reducing the interference from a high-rate user since the receiver and transmitter antenna weights can track precisely the changes in the DOAs of the desired signal.

VIII. Conclusion

This paper reviewed the channel structure and spreading code assignment in the physical layer, transport channel multiplexing into a physical channel along with sophisticated rate matching, and the key technologies of W-CDMA wireless access. Experimental results of the three-step cell search method associated with TSTD exhibited fast cell search time performance, while flexible continuous system deployment from outdoors to indoors was possible. Pilot symbol based coherent Rake receiver exhibited efficient, i.e., lower required E_b/I_0, transmission together with fast TPC and space diversity reception based on experiments in actual multipath fading channels. We also verified the effectiveness of site diversity and transmit diversity in decreasing the transmit power, thereby increasing link capacity. These experimental results associated with the successful system level experiments showed the superiority of W-CDMA wireless access technologies for commercial services.

Link capacity enhancing techniques such as those employing a COMSIC and CAAAD receiver/AAA-TD and examples of BS transceiver using these techniques were also presented with experimental results. The laboratory and field experimental results clarified the potential of the COMSIC and CAAAD receiver to reduce the mobile transmit power in the reverse link and that of AAA-TD to decrease MPI with large transmit power in the forward link. These techniques are essential for achieving high-speed packet transmission based on the W-CDMA air interface in an actual severe multipath fading channel.

Acknowledgements

The authors thank K. Okawa, S. Fukumoto, A. Morimoto, H. Taoka, and A. Harada of the Wireless Access Laboratory for their contributions to the W-CDMA experiments.

REFERENCES

[1] ADACHI (F.), SAWAHASHI (M.), and SUDA (H.), "Wideband DS-CDMA for Next Generation Mobile Communication System," *IEEE Commun. Mag.*, Vol. 36, pp. 56-69, (Sept. 1998).

[2] DAHLMAN (E.), GUDMUNDSON (B.), NILSSON (M.), SKOLD (J.), "UMTS/IMT-2000 Based on Wideband CDMA," *IEEE Commun. Mag.*, Vol. 36, pp. 70-80, (Sept. 1998).

[3] SAWAHASHI (M.), HIGUCHI (K.), TANAKA (S.), and ADACHI (F.), "Enhanced Wireless Access Technologies and Experiments for W-CDMA Communications," *IEEE Personal Communications,* vol. 7, n° 6, pp. 6-17, (Dec. 2000).

[4] 3GPP RAN, 3G TS 25.301 V3.5.0, (June 2000).

[5] 3GPP RAN, 3G TS 25.211 V3.4.0, (Sept. 2000).

[6] 3GPP RAN, 3G TS 25.213 V3.3.0, (Sept. 2000).

[7] OKAWA (K.) and ADACHI (F.), "Orthogonal Forward Link Using Orthogonal Multi-spreading Factor Codes for Coherent DS-CDMA Mobile Radio," *IEICE Trans. Commun.,* Vol. E81-B, pp. 777-784, (April 1998).

[8] 3GPP RAN, 3G TS 25.302 V3.5.0, (June 2000).

[9] 3GPP RAN, 3G TS 25.212 V3.4.0, (Sept. 2000).

[10] BERROU (C.), GLAVIEUX (A.), and THITIMAJSHIMA (P.), "Near Shannon Limit Error-correcting Coding and Decoding: Turbo-codes," *Proc. IEEE ICC'93*, pp. 1064-1070, Geneva, (May 1993).

[11] HIGUCHI (K.), SAWAHASHI (M.), and ADACHI (F.), "Fast Cell Search Algorithm in DS-CDMA Mobile Radio using Long Spreading Codes," *IEICE Trans.Commun.*, Vol. E81-B, pp. 1527-1534, (July 1998).

[12] HIGUCHI (K.), HANADA (Y.), SAWAHASHI (M.), and ADACHI (F.), "Experimental Evaluation of 3-Step Cell Search Method in W-CDMA Mobile Radio," P*roc. VTC '2000*, Tokyo, Japan, pp. 1527-1534, (May 2000).

[13] HIGUCHI (K.), HANADA (Y.), SAWAHASHI (M.), and ADACHI (F.), "Experiments on 3-Step Fast Cell Search Method using TSTD in W-CDMA Mobile Ratio," *IEICE Technical Report, RCS99-155*, pp. 99-105, (in Japanese), (Nov. 1999).

[14] ARIYAVISITAKUL (S.), "Signal and Interference Statistics of a CDMA System with Feedback Power Control-part II," *IEEE Trans. Commun.*, Vol. 42, pp. 597-605, (Feb./March/April 1994).

[15] SEO (S.), DOHI (T.), and ADACHI (F.), "SIR-Based Transmit Power Control of Reverse Link for Coherent DS-CDMA Mobile Radio," *IEICE Trans. Commun.,* Vol. E81-B n° 7, pp. 1508-1516, (July 1998).

[16] LING (F.), "Coherent Detection with Reference-symbol Based Estimation for Direct Sequence CDMA Uplink Communications," *Proc. VTC' 93*, New Jersey, pp. 400-403, (May 1993).

[17] ANDOH (H.), SAWAHASHI (M.), and ADACHI (F.), "Channel Estimation Filter Using Time-multiplexed Pilot Channel for Coherent Rake Combining in DS-CDMA Mobile Radio," *IEICE Trans. Commun.*, Vol. E81-B n° 7, pp. 1517-1526, (July 1998).

[18] ABETA (S.), SAWAHASHI (M.), and ADACHI (F.), "Adaptive Channel Estimation for Coherent DSCDMA Mobile Radio Using Time-Multiplexed Pilot and Parallel Pilot Structure" *IEICE Trans. Commun.*, Vol. E82-B n° 9, pp. 1505-1513, (Sept. 1999).

[19] HIGUCHI (K.), ANDOH (H.), OKAWA (K.), SAWAHASHI (M.), and ADACHI (F.), "Experimental Evaluation of Combined Effect of Coherent Rake Combining and SIR-based Fast Transmit Power Control for Reverse Link of DS-CDMA Mobile Radio," *IEEE J.Select.Areas Commun.*, Vo.18, pp.1526-1535, (Aug. 2000).

[20] HIGUCHI (K.), IKEDA (T.), FUKUMOTO (S.), SAWAHASHI (M.), and ADACHI (F.), "Experimental Evaluations of High Rate Data Transmission Using Turbo/Convolutional Coding in W-CDMA Mobile Radio" *Proc. Wireless 2000,* Calgary, pp. 687-693, (July 2000).

[21] SHIBUTANI (A.), SUDA (H.), and ADACHI (F.), "Multi-stage Interleaver for Turbo Codes in DS-CDMA Mobile Radio," *Proc. IEEE APCC/ICC*'98, pp. 391-395, (Nov. 1998).

[22] SHIBUTANI (A.), SUDA (H.), and YAMAO (Y.), "Performance of W-CDMA Mobile Radio with Turbo Codes Using Prime Interleaver," *Proc. VTC '2000,* Tokyo, Japan, pp. 1570-1574, (May 2000).

[23] GILHOUSEN (K.S), JACOBS (I.M), PADOVANI (R.), VITERBI (A.J), WEAVER (L.A), and WHEATLEY III (C.E.), "On the Capacity of a Cellular CDMA System," *IEEE Trans. Veh. Technol.*, Vol. VT – 40, n° 5, pp. 303-312, (May 1991).

[24 WONG (D.) and LIM (T.J), "Soft Handoffs in CDMA Mobile Systems," *IEEE Personal Commun. Mag.*, Vol. 36, pp. 6-17, (Dec. 1997).

[25] TIA/EIA/IS-95, "Mobile Station-Base Station Compatibility Standard for Dual-Mode Wideband Spread Spectrum Cellular System," Telecommunication Industry Association, (July 1993).

[26] FUKUMOTO (S.), HIGUCHI (K.), MORIMOTO (A.), SAWAHASHI (M.), and ADACHI (F.), "Combined Effect of Site Diversity and Fast Transmit Power Control in W-CDMA Mobile Radio," *Proc. VTC '2000*, Tokyo, Japan, pp. 1527-1534, (May 2000).

[27] MORIMOTO (A.), HIGUCHI (K.), FUKUMOTO (S.), SAWAHASHI (M.), and ADACHI (F.), "Experiments on Inter-Cell Site Diversity Using Two-step Selection Combining in W-CDMA Reverse Link," *IEICE Trans. Commun.*, Vol. E84-B n° 3, pp. 435-445, (March 2001).

[28] NEC, "Adjustment Loop in Down Link Power Control During Handover," TSG-RAN Working 1 Meeting, (Oct. 1999).

[29] Nortel, "Down Link Power Control During Soft Handover," 3GPP TSG-RAN WGI Meeting, (July 1999).

[30] FURUKAWA (H.), HAMABE (K.), and USHIROKAWA (A.), "SSDT– Site Selection Diversity Transmission Power Control for CDMA Forward Link," *IEEE J.Select.Areas Commun.,* Vol. 18, pp. 1546-1554, (Aug. 2000).

[31] HOTTINEN (A.) and WICHMAN (R.), "Transmit Diversity by Antenna Selection in CDMA Downlink," *Proc. IEEE ISSTA'98*, pp. 767-770, (Sept. 1998).

[32] FUKUMOTO (S.), SAWAHASHI (M.), and ADACHI (F.), "Performance Comparison of Forward Link Transmit Diversity Techniques for W-CDMA Mobile Radio," Proc. PIMRC'99, pp. 1139-1143, (Sept. 1998).

[33] 3GPP RAN, 3G TS 25.214 V3.4.0, (Sept. 2000).

[34] ALAMOUTI (S.M), "A Simple Transmit Diversity Technique for Wireless Communications," IEEE *J. Sel. Areas Commun.*, Vol. 16, n° 8, pp. 1451-1458, (Oct. 1998).

[35] FUKUMOTO (S.), HIGUCHI (K.), SAWAHASHI (M.), and ADACHI (F.), "Experiments on Space Time Block Coding Transmit Diversity (STTD) in W-CDMA Forward Link," Submitted to IEEICE *Trans. Fundamentals. Special Issue on Spread Spectrum*.

[36] DUEL-HALLEN (A.), HOLTZMAN (J.), and ZVONAR (Z.), "Multiuser Detection for CDMA Systems," IEEE *Personal Communications*, pp. 46-58, (April 1995).

[37] MOSHAVI (S.), "Multi-user Detection for DS-CDMA Communications," IEEE *Commun. Mag.*, pp. 124-136, (Oct. 1996).

[38] TSOULOS (G.V.), BEACH (M.A.), and MCGEEHAN (J.), "Wireless personal communications for the 21st century: European technological advances in adaptive antennas," IEEE *Commun. Mag.*, pp. 102-109, (Sept. 1997).

[39] COMPTON (R.T.), Jr., "An Adaptive Antenna in a Spread-spectrum Communication System," *Proc.* IEEE, Vol. 66, n° 3, pp. 289-295, (March 1978).

[40] HARADA (A.), TANAKA (S.), SAWAHASHI (M.), and ADACHI (F.), "Performance of Adaptive Antenna Array Diversity Transmitter for W-CDMA Forward Link," *IEEE PIMRC'99*, pp. 1134-1138, Osaka, (Sept. 12-15 1999).

[41] TANAKA (S.), HARADA (A.), SAWAHASHI (M.), and ADACHI (F.), "Transmit Diversity Based on Adaptive Antenna Array for W-CDMA Forward Link," *Proc. CDMA International Conference (CIC)'99*, pp. 282-286, Korea, (Sept. 1999).

[42] HARADA (A.), TANAKA (S.), SAWAHASHI (M.), and ADACHI (F.), "Experiments on Adaptive Antenna Array Transmit Diversity in W-CDMA Forward Link," *Proc. CDMA International Conference (CIC)'2000*, pp. 47-51, Korea, (Nov. 2000).

[43] TAOKA (H.), TANAKA (S.), IHARA (T.), and SAWAHASHI (M.), "Experiments on Adaptive Antenna Array Transmit Diversity with Carrier Frequency Calibration in Transmit Power-Controlled Forward Link for W-CDMA Mobile Radio," *Proc. 3G Wireless and Beyond*, San Francisco, (May 2001).

[44] TANAKA (S.), SAWAHASHI (M.), and ADACHI (F.), "Pilot Symbol-assisted Decision-directed Coherent Adaptive Array Diversity for DS-CDMA Mobile Radio Reverse Link," IEICE *Trans. Fundamentals,* Vol. E80-A, pp. 2445-2454, (Dec. 1997).

[45] TANAKA (S.), HARADA (A.), SAWAHASHI (M.), and ADACHI (F.), "Experiments on Coherent Adaptive Antenna Array Diversity for Wideband DS-CDMA Mobile Radio," IEEE *J.Select.Areas Commun.,* Vol. 18, pp. 1495-1504, (Aug. 2000).

[46] SAWAHASHI (M.), MIKI (Y.), ANDO (H.), and HIGUCHI (K.), "Pilot Symbol-Assisted Coherent Multistage Interference Canceller Using Recursive Channel Estimation for DS-CDMA Mobile Radio," IEEICE *Trans. Commun.*, Vol. E79-B, pp. 1262-1270, (Sept. 1996).

[47] SAWAHASHI (M.), ANDOH (H.), and HIGUCHI (K.), "Interference Replica Weight Control for Pilot Symbol-Assisted Coherent Multistage Interference Canceller Using Recursive Channel Estimation in DS-CDMA Mobile Radio," IEICE *Trans. Fundamentals,* Vol. E81-A, pp. 957-972, (May 1998).

[48] OKAWA (K.), HIGUCHI (K.), and SAWAHASHI (M.), "Parallel-type Coherent Multi-stage Interference Canceller with Iterative Channel Estimation Using Both Pilot and Decision-feedback Data Symbols for W-CDMA Mobile Radio," *IEICE Trans. Commun.*, Vol. E84-B n° 3, pp. 446-456, (March 2001).

[49] KOZONO (S.) and SAKAGAMI (S.), "Correlation coefficient on base station diversity for land mobile communication system," *IEICE Trans Commun.*, Vol. J70-B, pp. 476-482, (in Japanese), (Apr. 1987).

[50] MARTIN (U.), "Spatio-temporal radio channel characteristics in urban macrocells," *IEEE Proc.– Radar, Sonar Navig.*, Vol. 145, (Feb. 1998).

[51] IMT-2000 Study Committee, AIF/SWG2-25-4, (July 1998).

Chapter 3

Multicarrier CDMA techniques for future wideband wireless networks

M. Hélard and R. Le Gouable
France Télécom R&D
J.-F. Hélard and J.-Y. Baudais
INSA/LCST, France

I. Introduction

Future multimedia services (Internet, video transmission, data transfer, ...) will require the transmission of very high data rates over broadband radio channels. In order to meet the quality of service (QOS) requirements of these future multimedia applications, the multiple access techniques of next generation cellular systems should offer flexibility and especially very high spectral efficiency.

Spread spectrum has been very successfully used by the military services for decades and more recently has taken on a significant role in cellular and personal communications [1, 2]. Advantages of spread spectrum techniques are widely known: immunity against multipath distortion and jamming, low transmitted power, no need for frequency planning and last but not least, Code-Division Multiple Access (CDMA) capabilities [3]. As a result of the discussion in the last decade, CDMA will play an important role in the next generation of cellular systems, i.e. in UMTS and in IMT-2000 [4]. Thus, Wideband CDMA is a leading candidate for third generation wireless access. Based on direct sequence spread spectrum with a chip rate of 3.84 Mchip/sec, the transmitted signal occupies a bandwidth of about 5 MHz. It supports circuit and packet data access at nominal peak data rate equal to 384 Kbit/sec for macro cellular environments and to 2 Mbit/sec for indoor environments. However, we already know that wireless Internet usage is likely to be downlink limited. Indeed, for data services, peak bit rate is very important in determining overall system performance because of the highly bursty nature of Internet traffic.

On the other hand, the multi-carrier technique, well known under the acronym OFDM, which stands for Orthogonal Frequency Division Multiplexing, has been receiving widespread interest for wireless broadband multimedia applications over the last decade. The main advantages of this technique are its robustness in the case of frequency selective fading channels, its capability of portable and mobile reception and its flexibility. With this technique, the serial high rate baseband stream is distributed over many closely spaced orthogonal subcarriers. Due to the large number of subcarriers, the symbol duration is much higher than the channel time dispersion, which minimizes the Inter-Symbol Interference (ISI). Furthermore, residual ISI is suppressed thanks to the insertion of a guard interval larger than the channel time delay at the beginning of each symbol.

The OFDM technique was initially selected in 1990 in the European Digital Audio Broadcasting (DAB) standard [5, 6]. Its further prominent successes were in 1995 and 1998 as it was selected for the European Digital Video Broadcasting (DVB-T) [7] and in three broadband wireless indoor standards, namely ETSI-HIPERLAN/2 [8], American IEEE-802.11a [9] and Japanese MMAC. Furthermore, OFDM is becoming the basis of a world standard for asymmetric digital subscriber lines under the acronym of DMT for Digital Multitone [10].

The advantages and success of multi-carrier modulation and spread spectrum technique motivated many researchers to investigate the suitability of the combination of both techniques. This combination known as Multi-Carrier-Spread-Spectrum (MC-SS) benefits from the main advantages of both schemes: high spectral efficiency, high flexibility, multiple access capabilities, narrow-band interference rejection, simple one-tap equalization, etc., ... Three different concepts based on the combination of multi-carrier modulation with DS-CDMA have been introduced in 1993. They are known as "Multi-Carrier CDMA" (MC-CDMA) also called OFDM/CDMA [11–14], Multi-Carrier Direct Sequence CDMA (MC-DS-CDMA) [15] and Multi-Tone CDMA (MT-CDMA) [16]. The main differences between them are in the spreading that is carried out either in time or in frequency domain, in the frequency mapping and in the detection techniques.

Because of the limited frequency bandwidth on the one hand and the limited power of mobile units on the other hand, spectral and power efficiency of future systems should be as high as possible. This explains why the topic of MC-SS has recently received widespread interest from researchers, judging from the number of papers at international conferences. During the last years, deep system analysis and comparison of MC-SS and mainly MC-CDMA with DS-CDMA have been performed demonstrating the superiority of MC-CDMA

[17–19].The main topics studied are coding and modulation aspects, multi-user detection techniques as interference cancellation strategies, synchronization and channel estimation aspects, realization and implementation issues.

This paper is mainly devoted to the study of MC-CDMA systems for the downlink of high rate cellular networks over UMTS channels. The performance of different mono-user and multi-user detection techniques are compared. It is shown that a new linear detection technique named Global Minimum Mean Square Algorithm (GMMSE) offers very good performance mainly for non-full load systems. Furthermore, the application of channel coding as convolutional and turbo codes in synchronous MC-CDMA systems, in conjunction with this GMMSE detection technique is examined.

The paper is organized as follows: in section II, the three types of MC-SS schemes are reviewed. Their advantages and disadvantages in terms of transmitter and receiver structures, spectral efficiency and power efficiency are discussed. Then, in the following sections, the performance of MC-CDMA systems is studied in the case of a downlink transmission. Section III deals with analytical expressions to evaluate the performance of different detection techniques such as Maximum Ratio Combining (MRC), Equal Gain Combining (EGC) and Minimum Mean Square Error (MMSE). Monte Carlo simulations confirm the analytical results on a Rayleigh channel. After a brief description of the interference cancellation techniques in section IV, a new Global Minimum Mean Square Error (GMMSE) detection technique recently published in [20] is presented in section V and compared to the Parallel Interference Cancellation detection scheme. The UMTS channel modelling and system parameters used for simulations are described in section VI. Different results are given for a synchronous MC-CDMA system with various detection techniques, in order to look for the system that will offer the best trade-off between performance and complexity. With regard to the channel coding, convolutional codes and turbo-codes are considered. Finally, section VII summarizes the results and draws together the conclusions.

II. Multi-carrier spread spectrum concepts

II.1. MC-CDMA

The first concept, known as OFDM/CDMA or more often as MC-CDMA, is based on a serial concatenation of Direct Sequence spreading with Multi-Carrier Modulation. The MC-CDMA transmitter spreads the original data stream over different subcarriers in the frequency domain using a given spreading code. The

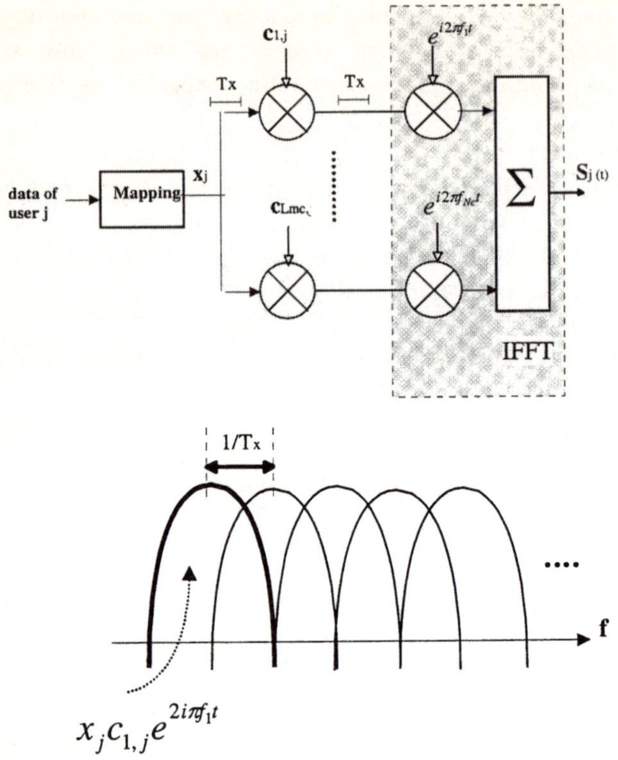

Figure 1.
MC-CDMA transmitter scheme and power spectrum of the transmitted signal.

effect of spreading is that different users can have access to the same carriers in a CDMA manner. The separation of the user's signals is then performed in the code domain. The advantage of MC-CDMA in comparison with DS-CDMA is that the spreading can be adapted to the frequency selective behaviour of the channel. Simple methods for signal detection in the frequency domain as one-tap equalizer per carrier can be used.

Figure 1 shows the MC-CDMA transmitter of the jth user and the power spectrum of the transmitted signal.

The data symbol $x_j(t)$ of the user j is transmitted in parallel over N_c subcarriers, each multiplied by one chip $c_{k,j}$ of the spreading code $C_j(t) = [c_{1,j}, c_{2,j}, \ldots, c_{Lmc,j}]$ assigned to user j. In this figure, the length L_{mc} of the spreading code is equal to the number N_c of subcarriers but this is not mandatory. As a consequence, the MC-CDMA systems offer an additional degree of freedom, and actually the number N_c of subcarriers is chosen to guarantee frequency non-selective fading over each subcarrier.

The expression of the transmitted signal $S_j\,(t)$ of user j during the time interval $[0, T_x]$ is:

$$(1) \quad S_j\,(t) = Re\left\{\sum_{k=1}^{N_c} x_j P\,(t) C_{k,j} e^{i2\pi f_k t}\right\},\ 0 \leq t < T_x$$

with $f_k = f_0 + \dfrac{k}{T_x}$

where T_x is the data symbol duration which is in this case equal to the OFDM symbol duration, x_j the data symbol transmitted during the signalling interval $[0, T_x]$, $P\,(t)$ the pulse shaping waveform which is generally rectangular and f_0 is the carrier frequency.

Practically, the MultiCarrier modulation and demodulation is easily carried out in the digital domain by performing IFFT and FFT operations. Furthermore the insertion between adjacent MultiCarrier (MC) symbols of a guard interval Δ, longer than the delay spread of the impulse response of the channel, guarantees the absence of Inter Symbol Interference (ISI). In this case, the resulting spectrum of the transmitted signal satisfies the orthogonality condition with the minimum frequency separation equal to $1/T_x = 1/(T'_x - \Delta)$, where T'_x is the new MC symbol duration and T_x is the useful duration of this symbol. In the receiver, after direct FFT and possibly de-interleaving, the received sequence is "equalized" in the frequency domain. Therefore, the MC-CDMA receiver can always employ all the received signal energy spread in the frequency domain. Undoubtedly, this is the main advantage of the MC-CDMA scheme compared to a DS-CDMA Rake receiver that has difficulties in making full use of the received signal energy scattered in the time domain.

For a synchronous system as the downlink mobile radio communication channel, the application of orthogonal codes such as Walsh-Hadamard codes guarantees the absence of Multiple Access Interference (MAI) in a gaussian channel. However, in non-ideal channels with frequency selective fading due to multipath propagation, the orthogonality between the signals of the different users is lost and MAI occurs. To combat the channel fading and thus the MAI, a multitude of detection techniques was proposed. They can be classified as either single-user detection (SD) or multi-user detection (MD). Single-user detection is performed by one tap equalization to compensate for the phase and amplitude distortions caused by the mobile radio channel. The one tap equalizer is simply one complex-valued multiplication per subcarrier. Basic SD techniques are Maximum Ratio Combining (MRC), Equal Gain Combining (EGC), Orthogonal Restoring Combining (ORC) or Minimum Mean Square Error (MMSE) equalization. This last technique, based on the MMSE criterion applied independently to each subcarrier achieves better performance [21, 25] as we will see in section III. Another way to mitigate the MAI is to optimize the spreading sequence selection within a given spreading sequence family as presented in [26].

With the aim to improve the performance of the receiver still further, Multi-user Detection (MD) can be processed, where the a priori knowledge about the spreading codes of the interfering users is exploited in the detection process. Indeed, the receiver jointly detects the other active user signals in order to mitigate the MAI. Multi-user detection methods can be divided into Interference Cancellation (IC) and Joint Detection (JD). The principle of IC is to detect the information of the interfering users and to reconstruct the interfering contribution in order to subtract it from the received signal. IC can be performed parallel for all interfering users with Parallel Interference Cancellation (PIC) detectors, or successively with Successive Interference Cancellation (SIC) detectors where only the strongest interferer remaining after the previous IC stage is cancelled.

The optimal detector applies Joint Detection (JD) with Maximum Likelihood Sequence Estimation (MLSE) or Maximum Likelihood Symbol-by-Symbol Estimation (MLSSE), respectively. Since the complexity of MLSE and MLSSE receivers grows exponentially with the number of users, their use is limited in practice to applications with a small number of users.

Most MC-CDMA systems were considered for optimum use of the available bandwidth of the synchronous downlink of a cellular network using orthogonal codes as Walsh-Hadamard codes. For the uplink, another scheme called Spread Spectrum Multi-Carrier Multiple Access (SS-MC-MA), which is derived from the MC-CDMA concept, has been proposed in [27]. In this case, each user k, exclusively transmits on a set of L subcarriers out of a total of N_c carriers.

II.2. MC-DS-CDMA

As shown in Figure 2, the MC-DS-CDMA transmitter modulates the substreams on subcarriers with a carrier spacing proportional to the inverse of the chip rate to guarantee the orthogonality between the spectrums of the substreams after DS spreading [15, 28].

First, the data stream is converted into parallel low rate substreams before applying the DS spreading on each substream in the time domain and modulating onto each subcarrier. This scheme was mainly investigated for an asynchronous uplink communication channel.

The transmitted signal $S_j(t)$ of user j is given by:

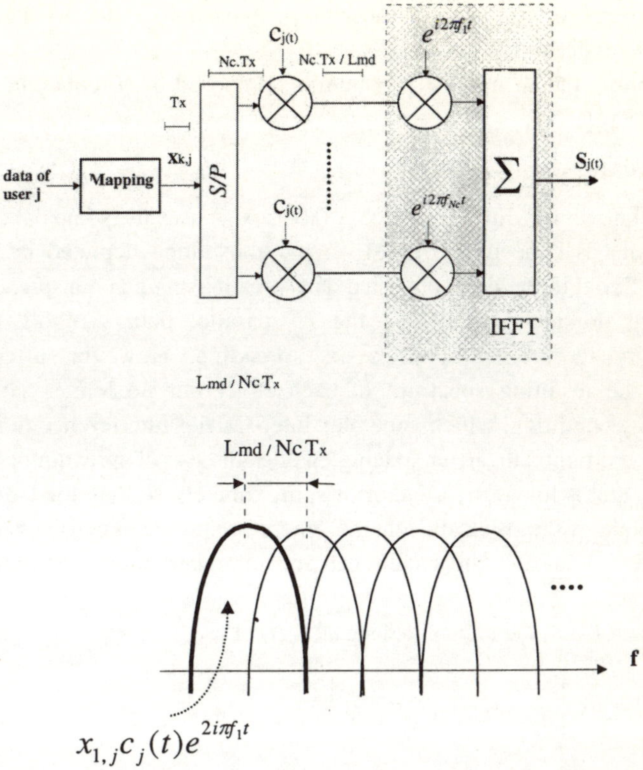

Figure 2.
MC-DS-CDMA transmitter scheme and power spectrumof the transmitted signal.

(2) $$S_j(t) = Re\left\{\sum_{k=1}^{N_c} x_{k,j} C_j(t) e^{i2\pi f_k t}\right\},$$

with $f_k = f_0 + \dfrac{k \cdot L_{md}}{N_c T_x}$

If the spreading code length L_{md} is inferior or equal to the number of subcarriers N_c, a single data symbol is spread in the time domain. By using a high number of subchannels, this concept benefits from time diversity. However, due to the frequency non-selective fading per subchannel, frequency diversity can only be exploited if channel coding with appropriate interleaving or subcarrier hopping is used, or if the same information is transmitted in parallel on several subcarriers [29]. In [30], a MC-DS-CDMA scheme with a subcarrier spacing larger than the inverse of the chip duration is proposed in order to increase the frequency diversity of the system.

In reception, the MC-DS-CDMA system is composed of N_c coherent receivers for data detection.

II.3. MT-CDMA

Concept 3, known as Multi-Tone-CDMA (MT-CDMA) uses the same data mapping and spreading as concept 2. The MT-CDMA transmitter, depicted in Figure 3, spreads the Serial to Parallel converted data stream using a given spreading code in the time domain [16]. Thus, the N_c parallel data symbols fulfil the orthogonality requirements before DS spreading. However, after the DS spreading, the resulting spectrum of each subcarrier no longer satisfies the orthogonality condition, which generates Inter-Carrier Interference (ICI). On the other hand, the tight subcarrier spacing enables the use of spreading codes with a length L_{mt} that is longer by a factor of approximately N_c than the length of the spreading code of the DS-CDMA scheme. Therefore, at the expense of higher ICI, under certain conditions, the system can accommodate more users than the DS-CDMA scheme.

The expression of the transmitted signal $S_j(t)$ of user j is:

$$(3) \quad S_j(t) = Re\left\{\sum_{k=1}^{N_c} x_{k,j} C_j(t) e^{i2\pi f_k t}\right\},$$

with $f_k = f_0 + \dfrac{k}{N_c T_x}$

which is identical to the expression of the MC-DS-CDMA transmitted signal given by equation (2) except that the subcarrier separation is equal to $1/(N_c T_x)$ instead of $L_{md}/(N_c T_x)$. The MT-CDMA receiver is composed of N_c Rake combiners, each of which has the same structure as the DS-CDMA Rake receiver. The MT-CDMA scheme suffers from inter-subcarrier interference, while the capability to use longer spreading codes, compared to the spreading codes assigned to a CDMA scheme, results in the reduction of Self Interference (SI) and MAI. In a channel where this improvement is dominant, the MT-CDMA scheme can outperform the DS-CDMA scheme [17]. This concept was mainly investigated for an asynchronous up-link scenario.

II.4. Comparison of different systems

In Table I, the main features of these three schemes are given in function of T_x, the input symbol duration.

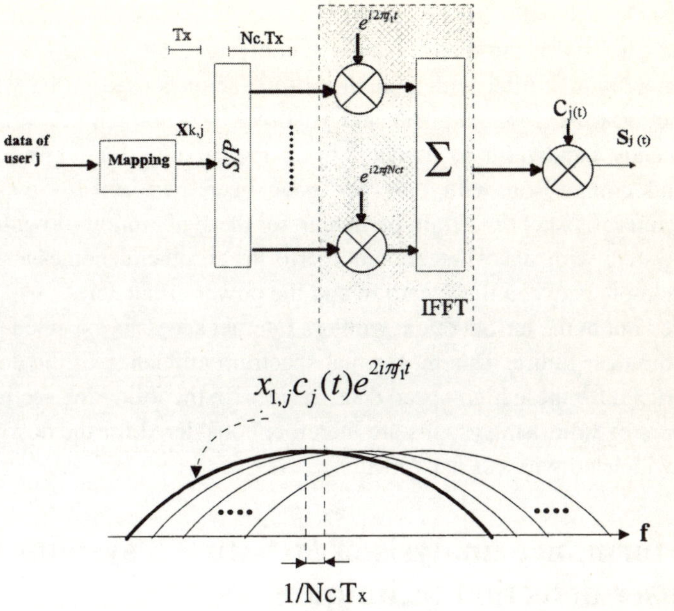

Figure 3.
MT-CDMA transmitter scheme and power spectrum of the transmitted signal.

Table I. *System features comparison.*

System	MC-CDMA	MC-DS-CDMA	MT-CDMA
Spreading	Frequency domain	Time domain	Time domain
Spreading length	L_{mc}	$L_{md} \approx L_{mc}$	$L_{mt} \approx N_c \,.\, L_{mc}$
Symbol duration at subcarrier	T_x	$N_c \,.\, T_x \,/\, L_{md}$	$N_c \,.\, Tx$
Subcarrier separation	$1 \,/\, T_x$	$L_{md} \,/\, (N_c \,.\, T_x)$	$1 \,/\, N_c \,.\, T_x$
Chip duration	T_x	$(N_c \,.\, T_x) \,/\, L_{md}$	$(N_c \,.\, T_x) \,/\, L_{mt}$
Required bandwidth (main lobe)	$(N_c + 1) \,/\, T_x$	$(N_c + 1)L_{md} \,/$ $(N_c \,.\, T_x)$	$0.5(2\, L_{mt} + N\text{-}1_c)$ $/(N_c.T_x)$ Roll-off factor = 0
Reception techniques	– MRC, EGC, ORC, MMSE – PIC, SIC, MLSE	- Coherent	– Rake combiner – MIMO equalizer
Main applications	– Downlink – Quasi- synchronous uplink	– Asynchronous – uplink	– Asynchronous plink

Since 1993, especially the first and the second schemes have been studied deeply. The capacity of the three systems is equivalent if $L_{mc} \approx L_{md} \approx (L_{mt}/N_c) \approx N_c$. When a Nyquist filter with a small roll-off factor is used in the MT-CDMA scheme, the MC-CDMA, MC-DS-CDMA and MT-CDMA systems require almost the same frequency bandwidth to transmit the same bit rate $1/T_x$. Deep system analysis and comparison with DS-CDMA have been performed showing the superiority of MC-CDMA [17–19]. In particular, for the synchronous downlink, the MC-CDMA system with MMSE detection outperforms all other schemes and offers the best trade-off between the spectrum and the power efficiencies.

As pointed out in the introduction, wireless Internet access is expected to grow rapidly in the near future. This means that spectrum efficiency of the downlink becomes crucial for these high-speed data services. In the following sections, the performances of MC-CDMA systems are therefore considered for the downlink of the future wideband wireless networks.

III. Performance analysis of MC-CDMA systems with single-user detection techniques

III.1. MC-CDMA transmitter and receiver

The block diagram of the MC-CDMA transmitter and receiver considered is depicted in Figure 4 for the downlink. Each data symbol x^n_j assigned to user j, $j = 1, ..., N_u$ and transmitted during the symbol interval n is multiplied with its user specific Walsh-Hadamard spreading code $C_j(t)=[c_{1,j}, c_{2,j}, ..., c_{Lmc,j}]^T$ of length L_{mc}, where $[.]^T$ denotes matrix transposition. L_{mc} corresponds to the

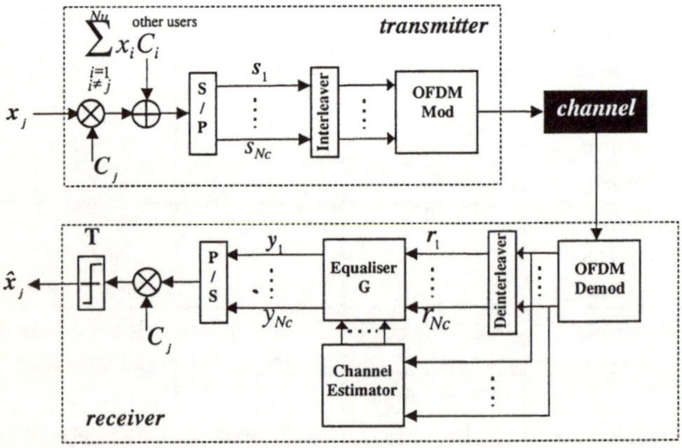

Figure 4. *Studied MC-CDMA transmitter and receiver.*

bandwidth expansion factor and is equal to the maximum number of simultaneous active users. The *j-th* column vector of the L_C x L_C matrix $\boldsymbol{C}$ corresponds to the spreading code $\boldsymbol{C}_j$ of the user j.

The vector of the data symbols transmitted during the *n-th* Orthogonal Frequency Division Multiplexing (OFDM) symbol by all the users can be written $\boldsymbol{X}^n = [x^n{}_1, x^n{}_2, \ldots, x^n{}_j, \ldots, x^n{}_{Lmc}]^T$, with $x^n{}_j = 0$ when user j is inactive.

Since we consider the synchronous downlink of an MC-CDMA system, the different data modulated spreading codes of the N_u users can be added before Serial-to-Parallel (S/P) conversion. Furthermore, the N_u user signals are supposed to be transmitted with the same power. The number N_c of subcarriers which are QPSK modulated is chosen equal to the spreading code length L_{mc}. For this study, frequency non-selective Rayleigh fading per subcarrier and time invariance during one OFDM symbol are assumed. The absence of Intersymbol Interference is also guaranteed by the use of a guard interval longer than the delay spread of the impulse response of the channel. Based on these assumptions and considering time and frequency interleaving, the complex channel fading coefficients are independent for each subcarrier and can be estimated for the subcarrier k by $h_k = \rho_k e^{i\theta_k}$. The signal received after the inverse OFDM operation (serial to parallel conversion and direct FFT) and de-interleaving can be expressed as:

$$(4) \qquad \boldsymbol{R} = [r_1, r_2, \ldots, r_{Nc}]^T = \boldsymbol{HCX} + \boldsymbol{N}$$

where the $N_c \times N_c$ diagonal matrix $\boldsymbol{H} = diag\{h_1, \ldots, h_{Nc}\}$ describes the complex channel frequency response and $\boldsymbol{N} = [n_1, n_2, \ldots, n_{Nc}]^T$ is the vector containing the Additive White Gaussian Noise (AWGN) terms with n_k representing the noise term at the subcarrier k with variance given by $\sigma_N^2 = E\{|n_k|^2\}, k = 1, \ldots, N_c$.

After equalization, the received signal can be written as:

$$(5) \qquad \boldsymbol{Y} = [y_1, y_2, \ldots, y_{Nc}]^T = \boldsymbol{GR} = \boldsymbol{GHCX} + \boldsymbol{GN}$$

In this section, the $N_c \times N_c$ matrix $\boldsymbol{G}$, which represents the complex equalization coefficients is diagonal with $\boldsymbol{G} = diag\{g_1, \ldots, g_{Nc}\}$. The different coefficients g_k can be derived from the channel estimation which is based on known transmitted pilot symbols inserted between the data carriers. Finally, after despreading and threshold detection, we obtain the detected data symbol $\boldsymbol{x}_j$, which corresponds to the sign of the scalar product of the received vector $\boldsymbol{Y}$ and the specific spreading code $\boldsymbol{C}_j$ as:

(6) $$\hat{x}_j = sign\,\langle Y, C_j\rangle = sign\left\{\underbrace{\sum_{k=1}^{N_c} c_{k,j}{}^2 g_k h_k x_j}_{\lambda} + \underbrace{\sum_{\substack{i=1\\ i\neq j}}^{N_u}\sum_{k=1}^{N_c} c_{k,i}C_{k,j}g_k h_k x_i}_{\lambda} + \underbrace{\sum_{k=1}^{N_c} c_{k,j}g_k n_k}_{\eta}\right\}$$

where λ represents the useful signal part, γ the MAI and η the noise term.

III.2. Single user detection systems: description and performance analysis

III.2.1. Description of different single user techniques

The calculation of the bit error probability P_b is based on the law of the large numbers, *i.e* the spreading code length L_{mc} and then the number of subcarriers N_c must be sufficiently high. In this case, owing to the fact that the different multiple access interference terms can be considered independent and using the central limit theorem, the interference term γ and the noise term η can be estimated by an additive zero mean Gaussian noise with a variance respectively equal to σ_γ^2 and σ_η^2. Furthermore, as the interference term and the noise term are independent, the sum of both has a zero mean Gaussian distribution with the variance equal to the sum of the variances. Therefore, the bit error probability can be estimated with:

(7) $$P_b = \frac{1}{2}\,erfc\left(\sqrt{\frac{E_\lambda}{2\,(\sigma_\gamma^2 + \sigma_\eta^2)}}\right)$$

where E_λ is the useful signal energy.

The average energy of the data bits of the different N_u user signals received is supposed to be identical and equal to E_b. Furthermore, thanks to the interleaving, the complex channel fading coefficients h_k are considered independent for each subcarrier and normalized such as $E[|h_k|^2] = 1$. Thus, the random distributed variable equal to the product of h_k with g_k has the first and second moment $E[g\,h]$ and $E[g^2\,h^2]$ respectively independent of the index k. So, when applying the law of large numbers, the useful in-phase signal energy is given by:

(8) $$E_\lambda = E[\Re\{g\,h\}]^2 E_b$$

In order to simplify the notation, we consider hereafter only the in-phase signal. Of course, the calculation is the same for the quadrature component. In the same way, the variance σ_γ^2 of the multiple access interference term for the in-phase signal is:

$$(9) \qquad \sigma_\gamma^2 = \sum_{\substack{i=1 \\ i \neq j}}^{N_U} \left(\sum_{k=1}^{N_c} c_{k,i} c_{k,j} \Re\{g_k h_k\} \right)^2 E_b$$

Since the codes are orthogonal, half of the products $c_{k,i} c_{k,j}$, $i \neq j$, are equal to $1/L_{mc}$ and the other half are equal to $-1/L_{mc}$. Then, the variance of the multiple access interference term can be written as:

$$(10) \qquad \sigma_\gamma^2 \approx \frac{N_u - 1}{L_{mc}} \left(\mathrm{E}\left[\Re\{gh\}^2 \right] - \mathrm{E}\left[\Re\{gh\} \right]^2 \right) E_b$$

and the variance of the additive zero mean Gaussian noise η is given by:

$$(11) \quad \sigma_\eta^2 = \mathrm{E}\left[\left(\sum_{k=1}^{N_c} c_{k,j} \Re\{g_k n_k\} \right) \right] \approx \mathrm{E}\left[g^2 \right] \frac{N_0}{2}$$

In the following, the values of E_λ, σ_γ^2 and σ_η^2 are presented for various single user detection techniques when the fading of the channel is Rayleigh distributed.

Maximum Ratio Combining technique (MRC)

In the single user case, MRC is the optimum diversity combining technique. The corresponding equalization coefficients are:

$$(12) \qquad g_k = h_k^*$$

where * stands for complex conjugation.

However, in a multi-user scenario, the multiplication by the conjugate complex channel coefficients results in enhanced MAI. The different quantities are equal to:

$$(13) \qquad E_\lambda = E_b$$

$$(14) \qquad \sigma_\gamma^2 + \sigma_\eta^2 = \frac{N_u - 1}{L_{mc}} E_b + \frac{N_0}{2}$$

Equal Gain Combining technique (EGC)

With EGC, only the phase shift is corrected. So, in fading channels the orthogonality of the Walsh-Hadamard spreading codes gets lost, resulting in MAI. The equalization coefficients are:

$$g_k = \frac{h_k^*}{|h_k|} \tag{15}$$

The variances of the multiple access interference term and the additive zero mean Gaussian noise are respectively:

$$E_\lambda = \frac{\pi}{4} E_b \tag{16}$$

$$\sigma_\gamma^2 + \sigma_\eta^2 = \frac{N_u - 1}{L_{mc}} \left(1 - \frac{\pi}{4}\right) E_b + \frac{N_0}{2} \tag{17}$$

Orthogonality Restoring Combining technique (ORC)

This technique, also called Zero Forcing (ZF), inverses the channel transfer function and thus restores the orthogonality between the users.

$$g_k = \frac{1}{h_k} \tag{18}$$

As the MAI is completely eliminated, performance does not depend on the number of active users. The drawback of ORC is that for small amplitude of h_k, the noise level is enhanced. For this detection technique, the analytic estimation of the bit error probability is not possible with this approach because the calculation of the variance σ_n^2 does not converge.

Minimum Mean Square Error technique (MMSE)

Among all these single-user detection techniques, MMSE equalization offers the best results. It minimizes the mean square value of the error ε_k between the signal s_k transmitted on subcarrier k and the assigned output y_k of the equalizer. The equalization coefficients based on this MMSE criterion applied independently per carrier are equal to:

$$g_k = \frac{h_k^*}{|h_k|^2 + \frac{1}{\zeta_c}} = \frac{h_k^*}{|h_k|^2 + \frac{N_c}{N_u}\frac{N_0}{2E_b}} \tag{19}$$

where ζ_c is the subcarrier signal to noise ratio.

To calculate the bit error probability, we set $z = \Re\{g_k h_k\}$ and $x = |h_k|^2$. As $E[|h_k|^2] = 1$, the probability density of x is $p(x) = e^{-x}$. Therefore, we obtain:

$$E_\lambda \approx E[z]^2 E_b \tag{20}$$

$$\sigma_\gamma^2 + \sigma_\eta^2 \approx \frac{N_u - 1}{L_{mc}} \left(E[z^2] - E[z]^2\right) E_b + E\left[|g_k|^2\right] \frac{N_0}{2} \tag{21}$$

with

$$E[z] = \int_0^{+\infty} \frac{xe^{-x}}{N_u x + N_c \dfrac{N_0}{2E_b}} dx, \tag{22}$$

$$E[z^2] = \int_0^{+\infty} \frac{x^2 e^{-x}}{\left(N_u x + N_c \dfrac{N_0}{2E_b}\right)^2} dx, \tag{23}$$

$$E\left[|g_k|^2\right] = \int_0^{+\infty} \frac{xe^{-x}}{\left(N_u x + N_c \dfrac{N_0}{2E_b}\right)^2} dx \tag{24}$$

As already reported in Section III.1, the matrix $\boldsymbol{G}$ is diagonal for all these basic single user detection techniques which means that the sequence received is equalized by using a bank of N_c adaptive one tap equalizers which results in a low complexity equalizer.

III.2.2. Analytical and simulation results with single user detection techniques

The analytical and simulation results with MRC, EGC and MMSE detections are presented in Figure 5 for a Rayleigh channel. The number N_u of active users is equal to the spreading code length L_{mc} (full user capacity) which is also equal to the number $N_c = 64$ of subcarriers.

The Matched Filter (MF) bound or "limit bound" is given as reference. It corresponds to the BER obtained in the case of MRC detection for a MC-CDMA system with only one user and consequently not affected by MAI.

It can be seen that the analytical results are really close to the simulation results, particularly for MRC and EGC detections. For MMSE detection, there is a

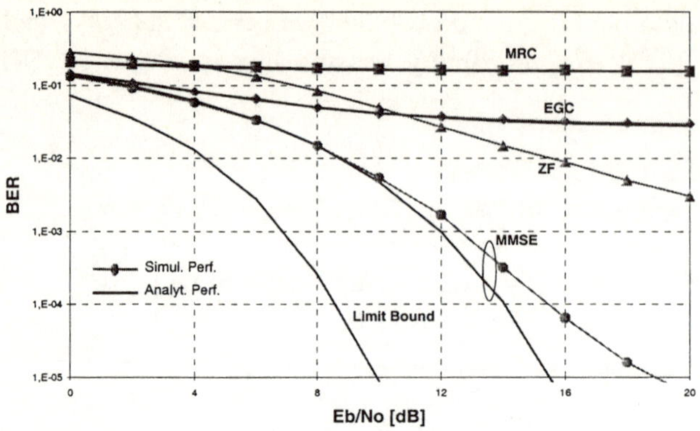

Figure 5. Single user detection techniques: analytical and simulation results; full load system: $N_u = L_{mc} = N_c = 64$.

slight difference between analytic and simulated results because the contribution of the multiple access interference is not exactly Gaussian. Indeed, for the analytic approach, the number of interfering terms is not large enough to obtain a Gaussian resulting signal according to the central limit theorem. Nevertheless, the MMSE outperforms the other techniques avoiding an excessive noise amplification for low signal to noise ratios while restoring the orthogonality among users for large signal to noise ratios. The potential of MMSE already pointed out in many references [21, 25] is there confirmed. However, this MMSE equalization per carrier method is not optimal, since it does not take into account the despreading process and thus does not minimize the mean square error at the input of the threshold decoder. In order to obtain better performance, a new method based on a global implementation of the MMSE criterion is presented in Section V.

IV. Multi-user detection techniques

There are two major types of Multi-User (MU) detectors MU detectors based either on the Maximum Likelihood algorithm or on the Interference Cancellation (IC). They both need the knowledge of all the spreading codes used and consequently are more complex to implement than SU detectors. However, a significant gain is expected with the use of the MU detectors. The MLD (Maximum Likelihood Detector) (MLD) may be carried out by a Viterbi decoder. Nevertheless, the MLD detector has quite a high complexity that exponentially increases with the number of active users [31]. So, in order to handle a large number of users, receivers can implement suboptimal nonlinear interference cancellation (IC) techniques with lower complexity. We will describe the two types of IC detectors based on either a serial (SIC) or a parallel (PIC) interference cancellation [32, 33].

The aim of each of these IC detectors is to estimate the contribution of each user in the MAI term and to suppress it. The process may be performed iteratively in multiple detection stages. The useful received signal is first equalized by a SU technique and then despread by each code. Since the downlink is considered synchronous, an Inverse Fast Hadamard Transform (IFHT) may be carried out to reduce the receiver complexity.

IV.1. Successive Interference Cancellation

The SIC detector first detects the most powerful interfering user and then cancels its contribution from the received signal. The second strongest interferer is then cancelled and so on. The processing may be repeated for a few or for all users. A complete detector would consider all users, but only the interferers stronger than the useful one are commonly suppressed. SIC detector is generally used when the power of some users are higher than the power of the useful user. Since processing one supplementary stage leads to an additive time delay, a trade-off between the number of stages and the total acceptable delay has to be found. Figure 6 shows the scheme of one stage SIC detector where the strongest interfering user is detected among the estimations $\tilde{x}_1 \ldots \tilde{x}_{NU}$ and then cancelled from the received signal. The process is carried out iteratively until the remained interferers are considered insignificant. The resulting signal is finally despread. The data detection may be hard or soft.

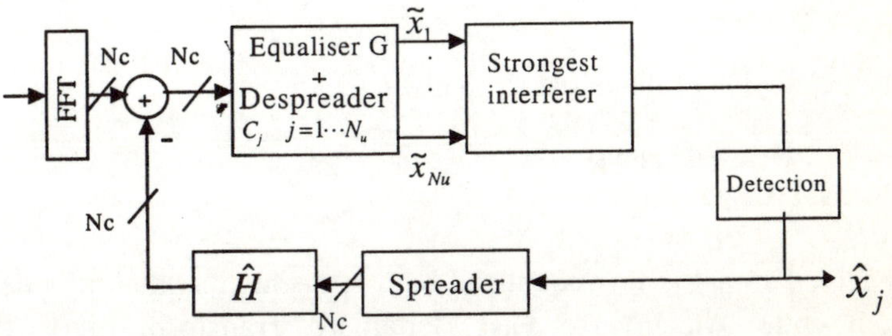

Figure 6. One stage SIC detector scheme.

IV.2. Parallel Interference Cancellation

The Parallel Interference Cancellation (PIC) structure is based on an estimation of the total interference due to the simultaneous other users in order to remove it from the received signal. Figure 7 shows the scheme of a PIC detector, where the contribution of all interfering users is cancelled in parallel, reducing the

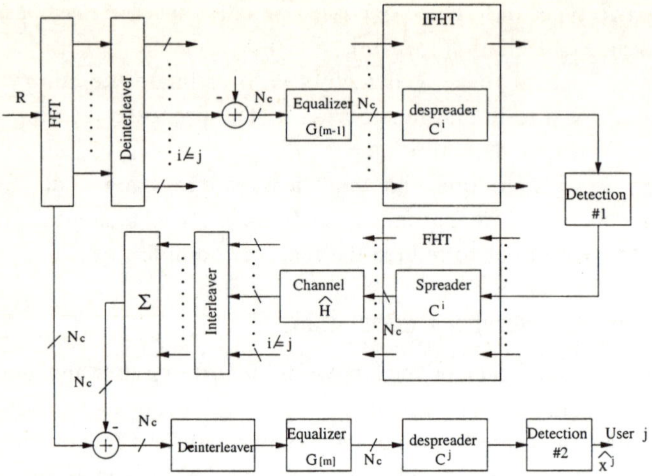

Figure 7. *One stage* PIC *detector scheme.*

time delay of a SIC detector. The expression of this iterative system for the *m-th* stage and the *j-th* user is given by the following:

$$(25)\quad \hat{x}_j^{[m]} = \mathbf{C}_j^{*T}\mathbf{G}^{[m]}\left(\mathbf{R} - \mathbf{H}\sum_{i=1,\, i \neq j}^{N_u} \hat{x}_i^{[m-1]}\mathbf{C}_{(i)}\right)$$

$m = 1, \ldots, M_{\text{it}}$

with the expression of the initial stage given by:

$$(26)\qquad \hat{x}_j^{[0]} = \mathbf{C}_j^{*T}\mathbf{G}^{[0]}\mathbf{R} \qquad\qquad j = 1, \ldots, N_u$$

The received signal is first equalized by a SU technique, then it is despread by each code. An Inverse Fast Hadamard Transform (IFHT) can be implemented since the system is synchronous. As for the SIC detector, data detection may be either hard or soft. After detection, the data is spread again, tapped by the estimated channel coefficients $\hat{H}$ and then subtracted from the signal received. Finally, the resulting signal with lower MAI term is then equalized, despread and detected. We can note that the second equalizer structure (G[m]) may be different from the first (G [m – 1]).

V. A new global minimum mean square error (GMMSE) detection technique

The aim of this new method, which has been patented [34] and named the Global MMSE algorithm, is to minimize the mean square error between the transmitted symbol x_j and the estimated symbol $\hat{x}_j$ [20]. Let $W_j^T = [w_j^0, w_j^1, ..., w_j^{Nc}]$ be the optimal weighting vector. The estimated symbol of the *j-th* user is:

$$\hat{x}_j = W_j^T R = C_j^T G R \quad (27)$$

According to Wiener filtering, the optimal weighting vector is:

$$W_j = \Gamma_{R,R}^{-1} \cdot \Gamma_{R,x_j} \quad (28)$$

where $\Gamma_{R,R}$ is the autocorrelation matrix of the received vector R and $\Gamma_{R,xj}$ is the cross-correlation vector between the desired symbol x_j and the received signal vector R. These terms are equal to:

$$\Gamma_{\mathbf{R},\mathbf{R}} = E\left\{\mathbf{R}^*\mathbf{R}^T\right\} = \mathbf{H}^*\mathbf{C}E\left\{\mathbf{X}^*\mathbf{X}^T\right\}\mathbf{C}^T\mathbf{H} + E\left\{\mathbf{N}^*\mathbf{N}^T\right\} \quad (29)$$

$$\Gamma_{\mathbf{R},x_j} = E\left\{\mathbf{R}^*x_j\right\} = \mathbf{H}^*\mathbf{C}E\left\{\mathbf{X}^*x_j\right\}$$

The optimal weighting vector can be written as:

$$\mathbf{W}_j^T = E\left\{x_j\mathbf{X}^{*T}\right\}\mathbf{C}^T\mathbf{H}^*\left(\mathbf{HC}E\left\{\mathbf{XX}^{*T}\right\}\mathbf{C}^T\mathbf{H}^* + E\left\{\mathbf{NN}^{*T}\right\}\right)^{-1} \quad (30)$$

The subcarrier noises have the same variance and are independent. Thus, $E\{NN^{*T}\} = \sigma_N^2.I$, where I is the identity matrix. Since the user signals have the same power ($E\{x_j^2\} = E_s$) and are independent, we can write $E\{XX^{*T}\} = E_s.A$, where $A = \{a_{ij}\}$ is a diagonal matrix with the term $a_{jj} = 1$ if the user j is active and $a_{jj} = 0$ if the user j is inactive. Then, the equalization coefficient matrix is:

$$\mathbf{G} = \mathbf{H}^*\left(\mathbf{HCAC}^T\mathbf{H}^* + \frac{\sigma_N^2}{E_s}\mathbf{I}\right) \quad (31)$$

In the full load case ($N_u = L_{mc}$) and only in that case, the quantity $C.A.C^T$ is equal to the identity matrix and the equalization coefficients matrix G is a diagonal matrix with the *k-th* subcarrier equalization coefficient equal to the former Equation (19).

On the other hand, when the capacity is not full ($N_u < L_{mc}$), the equalization coefficient matrix G is no more diagonal. In that case, as will see in the last section, the Global MMSE (GMMSE) algorithm outperforms the MMSE per carrier algorithm, since it minimizes the decision error taking into account the de-spreading process instead of minimizing the error independently on each subcarrier.

VI. Comparison of different detection schemes: simulation results

VI.1. System and simulation parameters

VI.1.1. System parameters

Results presented in this final part concern simulation results over channels that are representative of vehicular test environments in UMTS standard. For our study, we considered BU (Bad Urban) and HT (Hilly Terrain) channels as defined in GSM standard [35] and two UMTS channels defined for vehicular environments [36]. In our simulations, the propagation channels consist of 6 or 12 power loss paths and lognormal shadowing of zero mean with classical Jakes Doppler spectrum on each path. Table II describes the tapped-delay line parameters for BU, HT, VA and VB channels as defined by [35] and [36].

Table III gives the carrier frequency F_0, the maximum velocity V_m of the mobile user, the resulting maximum Doppler frequency f_{Dmax} and the delay spread T_m for each channel.

***Table II**. Tapped delay line parameters of BU, HT, VA and VB channels.*

BU: Bad Urban		HT: Hilly Terrain		Vehicular A		Vehicular B	
Delay μsec	Power dB	Delay μsec	Power dB	Delay μsec	Power dB	Delay μsec	Power dB
0.0	– 7	0.0	– 10	0.0	0	0.0	– 2.5
0.2	– 3	0.2	– 8	0.31	– 1	0.3	0.0
0.4	– 1	0.4	– 6	0.71	– 9	8.9	– 12.8
0.8	0	0.8	– 4	1.09	– 10	12.9	– 10.0
1.6	– 2	1.6	0	1.73	– 15	17.1	– 25.2
2.2	– 6	2.0	0	2.51	– 20	20.0	– 16.0
3.2	– 7	2.4	– 4				
5.0	– 1	15.0	– 8				
6.0	– 2	15.2	– 9				
7.2	– 7	15.8	– 10				
8.2	– 10	17.2	– 12				
10.0	– 15	20.0	– 14				

Table III. Channel and system parameters.

Channel	BU	HT	VA	VB
F_0	1.8 GHz	1.8 GHz	2 GHz	2 GHz
V_m	1 m/sec	27.7 m/sec	27.7 m/sec	27.7 m/sec
f_{Dmax}	6 Hz	166.7 Hz	185 Hz	185 Hz
T_m	10. µsec	20. µsec	2.5 µsec	20. µsec
Δ	20. µsec	20. µsec	5. µsec	20. µsec
T_x	128. µsec	128. µsec	33.33 µsec	133.33 µsec
N_c	256	256	256	1024
η	0.793 bit/sec/Hz	0.793 bit/sec/Hz	0.797 bit/sec/Hz	0.797 bit/sec/Hz
W	2 MHz	2 MHz	7.68 MHz	7.68 MHz

The last parameters of the table depending on the system are given for each channel:

– the guard interval Δ that is equal or longer than T_m,
– the useful OFDM symbol duration T_x,
– the number N_C of subcarriers,
– the total spectral efficiency η, given for a system including a 1/2 rate channel code,
– the bandwidth W of the transmitted signal.

The characteristics of the simulated MC-CDMA system are the following:

– the length L_{mc} of the Walsh Hadamard spreading code is equal to 64,
– different data per user per OFDM symbol for BU, HT and VA channels and 16 for VB channel,
– the maximum number of users N_u is equal to 64 when the system is full load,
– the subcarriers are QPSK modulated,
– a random frequency interleaving is carried out over one OFDM symbol,
– all users have the same power.

Channel estimation and synchronization are assumed to be perfect and based on the insertion of pilot symbols between the data. The arrangement of these pilots must guarantee an optimum sampling of the channel transfer function in time and in frequency, which depends on the coherence bandwidth and the coherence time of the channel. Nevertheless, in this case, we estimate that a good sampling of the channel response can be reached with the insertion of one pilot every N_p = 12 carriers, which induces a loss of spectral efficiency equal to 8.3%. So, as we use a channel code rate R = 1/2 associated to QPSK modulation (M = 2 bits per symbol), the total spectral efficiency is equal for a full load system to:

$$(32) \qquad \eta = \frac{N_p - 1}{N_p} \cdot M \cdot R \cdot \frac{T_x}{T_x + \Delta} = 0.917 \frac{T_x}{T_x + \Delta}$$

Furthermore, all users are considered to have the same power, which means that power control may be implemented. For the comparison of the different detection systems, Bit Error Rate (BER) performance is provided as a function of E_b/N_0 where E_b is the energy per useful bit. All results have been obtained through simulation with the Monte Carlo method. For most results, 2 limit curves are provided:

- a lower limit called "limit limit bound" that stands for a system with only one user and a matched filter (MRC detection) and consequently without MAI,
- a upper limit that corresponds to a "full load" system with the maximum MAI.

In order to evaluate the performance of the MC-CDMA system on the downlink, we first study the degradations due to MAI. For all systems including interference cancellation techniques, a one stage PIC detector with hard decision is implemented. Therefore, we first obtain a set of curves versus N_u the number of users. Then we introduce channel coding and analyse the behaviour of a complete system.

VI.2. Comparison between different detectors without channel coding

We will compare performance of 3 different detectors: one Single User (SU) detector, one Multi User (MU) detector and a GMMSE detector.

For performance of a system including a SU detector, only the results of the MMSE algorithm (Minimum Mean Square Error algorithm) are presented. Indeed, as we have already seen in Section III.2.1, the SU MMSE detector revealed to be the best single user detector [22]. For the MU detector, we only present results with one stage of cancellation PIC detector where the IC process is carried out with the MMSE technique. In fact, PIC and SIC detectors provide equivalent results for quite similar complexity. Carrying out a second stage of PIC does not largely improve performance; indeed, the gain remains under 1 dB for a higher complexity [23]. In order to evaluate the influence of MAI, we first present results without channel coding for most channels.

VI.2.1. Full load system performance

Figure 8 illustrates the BER performance of SU and MU detectors for a full load system corresponding to 64 users. The curves obviously demonstrate how

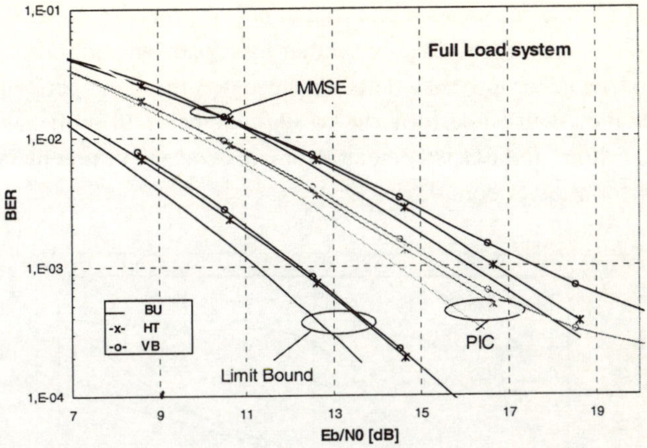

Figure 8. *Performance of the SU MMSE and MU PIC detectors for a full load system over BU, HT and VB channels.*

efficiently MAI is fought by a MU detection technique that widely outperforms the SU MMSE detector. Nevertheless, at BER=10^{-3}, a degradation of at least 2 dB remains, compared to the limit bound. Results with GMMSE are not presented in this figure because, as explained in Section V, they are exactly the same as those of SU MMSE technique for a full load system.

We also note slight differences depending on the channels. As expected, poorer performance is obtained over the VB channel – the most severe one. Over the VA channel, for which results are not represented, we obtained the same performance as over the HT channel. For the VB channel, we can note a floor effect due to the time selectivity of the channel when OFDM duration increases; in fact with only 256 subcarriers, which curves are not represented, there is no floor effect. This phenomenon will be cancelled by channel coding.

VI.2.2. System performance versus the capacity

The performance of the SU MMSE, MU PIC and GMMSE detectors are compared for systems with different number of users over BU channel. Figure 9 illustrates the performance of each detector in term of capacity related to the number of active users; the curves are drawn for a BER equal to 10^{-3} and consequently provide the necessary E_b/N_0 to reach this BER. As expected, performance improves when the number of users N_u decreases since the MAI level decreases. As for the full load system, MU techniques outperform SU techniques. The GMMSE curve tracks the MU PIC curve fairly closely up to a 3/4 load system, then the PIC detector outperforms the GMMSE detector. As mentioned in Section V, at full load, GMMSE and SU MMSE have the same performance.

The total bit rate increases with the number of users N_u. Nevertheless, most of time, systems are not full load but rather half load. Thus, in meantime, using either a PIC detector or a GMMSE detector presented the same performance up to 48 users and they both outperform the SU MMSE detector. If implemented with an adaptive algorithm, the GMMSE detector remains easier to implement than a MU PIC detector for a lower complexity [25].

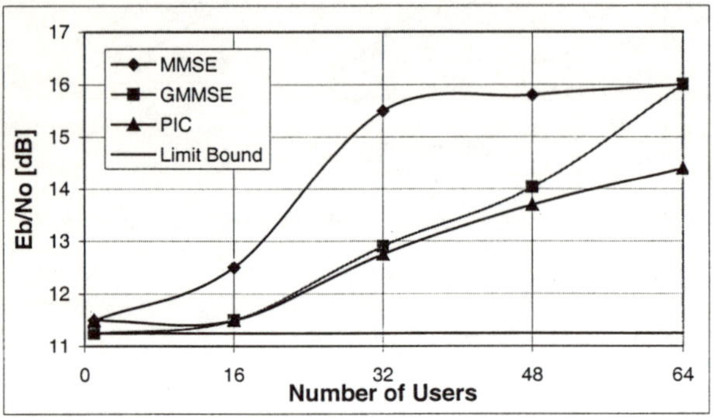

Figure 9. Performance versus capacity over a BU channel at BER =10^{-3}.

VI.3. Multiuser detector including channel coding

By introducing redundancy in the transmission system, channel encoding allows better performance, especially when channel fading, noise or interference occurs [37]. At the reception level, in order to achieve an optimal efficiency, the Viterbi decoder requires a reliability measurement of the data given by a Log-Likelihood Ratio (LLR) function. An estimation of the LLR function for MC-CDMA systems using QPSK modulation is given in [38]:

$$L_i = \frac{4}{\sigma^2} \frac{\sum_{k=1}^{N_c - 1} \rho_k Z_i}{N_c} \tag{33}$$

where L_i depends on the mean value of the channel fading over N_c subcarriers on which the transmitted data has been spread and z_i is the symbol obtained after despreading.

Last results include either a classical Convolutional Coding (constraint length K = 7, polynomial generators: 133o and 177o) or a Convolutional Turbo-Code (constraint length K = 5, polynomial generator: 23o and 35o). The two encoders are half rate. As mentioned in [23], decoding might be introduced into the IC process in order to obtain a better estimation of the interfering signals; nevertheless the complexity largely increases with the number of active users, limiting them to a small number. Thus, for our simulation, channel decoding is performed outside the iterative process for GMMSE and MU detectors.

VI.3.1. Influence of time interleaving

Figure 10 illustrates BER performance with convolutional coding for a full load system over HT channel. Curves respectively correspond to results obtained with PIC and SU MMSE detectors with and without time interleaver. We implemented a random 12 msec interleaver that corresponds to about twice the coherence time of the HT, VA and VB channels.

For results obtained without time interleaver, we can note on the 2 upper curves that PIC detector performance is worse than MMSE detector performance. This phenomenon, already pointed out by [32], is mainly due to bad decisions in the IC process which leads to bursts of errors at the output of the convolutional decoder.

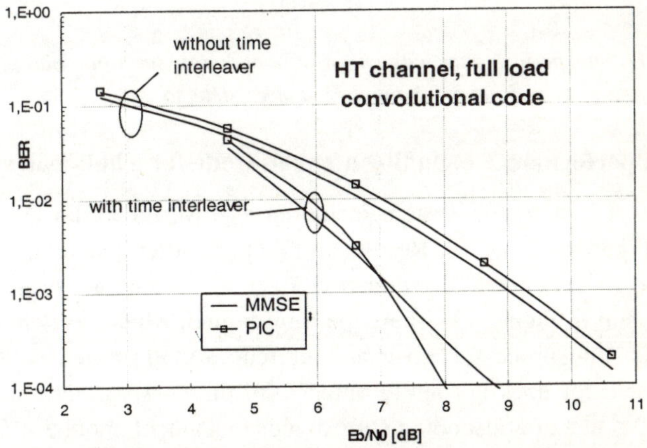

Figure 10. *Influence of interleaving for a full load system over a HT channel with convolutional code.*

Thanks to the time diversity of the channel, time interleaving provides a significant gain highlighted by results of Figure 10. Otherwise carrying out the time interleaving allows PIC detector to outperform the SU MMSE detector at about 10^{-3} as shown by the lower curves in Figure 10. Introducing time interleaving leads to a 3 dB gain for PIC detector and to a 2 dB gain for the MMSE detector at BER = 10^{-4}; these substantial gains confirm the opportunity of using a time interleaver for these selective channels.

Figure 11 illustrates the performance of an MMSE detector over HT, VA and VB channels with time interleaving for a full load system. We can note a 1.75 dB gain on performance over HT channel at BER=10^{-3} when introducing time interleaving, and about a 2 dB gain for channel VB. Over the VA channel that is less frequency and time selective than HT and VB channels, the coding gain is reduced to 1.5 dB.

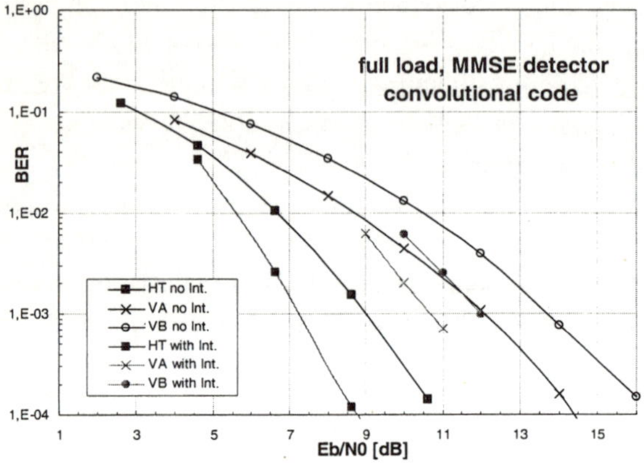

Figure 11. *BER performance with convolutional coding for a full load system, different channels and MMSE detector.*

VI.3.2. BER performance including a Turbo-Code for a half load system

Performance of a MC-CDMA system including a powerful Turbo Decoder [39] is given in Figures 12 and 13. Results are obtained after 4 iterations of Turbo decoding for both channels. Performances over VA channel are presented for a half load system in Figure 12, where the limit bound, which takes into account the Turbo Decoding process, is given as a reference. As illustrated by BER curves, the MU PIC detector does not really outperform the SU MMSE detector. In fact, reaching a 10^{-4} BER after decoding corresponds to a BER of about 5 $\times 10^{-2}$ before decoding where performance of all detectors are equivalent (cf. Figure 8). Thus, the interest of implementing either a SU or a MU IC technique will depend on their respective performance before decoding [40].

Figure 12 also illustrates the good performance of the GMMSE detector especially for a nonfull load system as already pointed out without channel coding in Figure 9, where the GMMSE detector had equivalent performance to the MU detector. Moreover, in the case of channel coding, the GMMSE algorithm widely outperforms the MU detector. In fact, the PIC detector is not optimal at low signal to noise ratios because of bad decisions taken in the estimation of the interferers in the IC process.

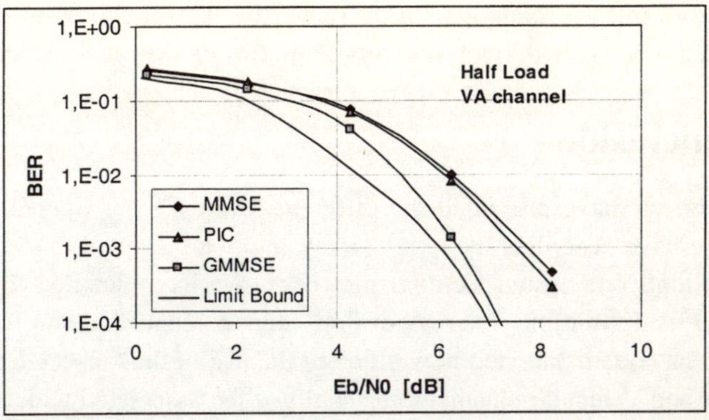

Figure 12. *BER performance for a half load system over a VA channel.*

VI.3.3. BER performance with a Turbo Decoder versus the number of active users

In Figure 13, performance of each detector over the BU channel is given versus the number of users. As in Figure 9, the curves are drawn for a BER equal to 10^{-3} and consequently provide the necessary E_b/N_0 to reach this BER.

For the BU channel, we did not carry out time interleaving. In fact, this channel would require a 333 msec interleaver in order to fight symbol correlation and implementing such a long interleaver is not possible for mobile applications. Nevertheless, we can note that up to 1/3 load the GMMSE detector allows performance close to the limit bound and is then always better than PIC and MMSE detectors whilst the PIC detector never widely outperforms SU MMSE. In fact, as already pointed by results in Figure 10, for a full load system, the MMSE detector even outperforms the PIC detector. The interest of implementing a

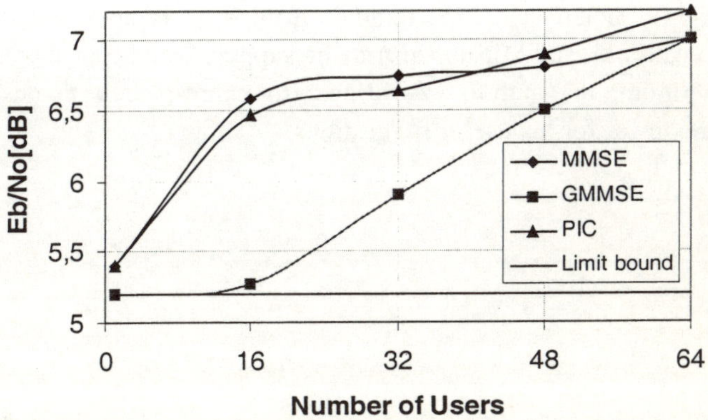

Figure 13. *Performance of a MC-CDMA system including a Turbo Decoder on a BU channel.*

GMMSE algorithm associated with a powerful Turbo Code is obviously demonstrated here since it always widely outperforms the other detectors.

VII. Conclusion

In this paper, we have presented the different Multi Carrier Spread Spectrum systems. We have described the MC-CDMA system well adapted to the downlink of wireless high rate system cellular networks. We have detailed the existing Single User and Multiple User algorithms and presented a new one called GMMSE. We have demonstrated how efficient the use of the GMMSE detector was over BU, HT and Vehicular channels, especially when associated with a powerful Turbo Code. Even if some system parameters may be optimized depending on the application, the performance of MC-CDMA systems presented in this paper are quite acceptable leading to a total spectral efficiency (including the losses due to the insertion of pilot symbols for channel estimation) of about 0.8 bit/(sec ·Hz) for each studied channel. This spectral efficiency corresponds to 0.0125 bit/(sec ·Hz) per user in our system where 64 users can simultaneously transmit. With respect to UMTS and IMT2000 requirements, it is possible to allocate to one user several or all user codes such that, for a bandwidth equal to 5 MHz, a net bit rate up to 4 Mbit/sec can be assigned to a single user. A 6 Mbit/sec bit rate could be reached with a 3/4 rate code. Thus, it is shown that MC-CDMA is a very promising multiple access scheme especially for the downlink of future mobile radio systems. Nevertheless, we have to keep in mind that some parameters such as channel estimation, nonlinearity amplifiers and phase noise oscillators, as well as synchronization devices have not yet been taken into account to complete this analysis.

Acknowledgments

The authors, J.-F. Hélard and J.-Y. Baudais from INSA, Rennes would like to thank FT R&D/DMR/DDH and Mitsubishi ITE who supported and contributed to this study. Furthermore, the authors would like to express their thanks to the anonymous reviewer for his useful suggestions.

REFERENCES

[1] TURIN (G.L.), Introduction to spread-spectrum antimultipath techniques and their application to urban digital radio, *Proc. of the IEEE* **68**, n° 3, pp. 328-353, (March 1980).

[2] PICKHOLTZ (R.L), MILSTEIN (L.B.), SCHILLING (D.L.), Spread spectrum for mobile communications, *IEEE Trans. Vehic. Tech.* , **40**, n° 2, pp. 313-322, (May 1991).

[3] LEE (W.C.Y.), Overview of cellular CDMA, *IEEE Trans. Vehic. Tech.*, **40**, n° 2, pp. 291-302, (May 1991).

[4] SHAFI (M.), Wireless communication in the twenty-first century: a perspective, *Proceedings of the IEEE*, **85**, n° 10, pp. 1622-1638, (1997).

[5] ALARD (M.), LASSALE (R.), Principles of modulation and channel coding for digital broadcasting for mobile receivers, *EBU review*, **224**, pp. 3-25, (Aug. 1987).

[6] LE FLOCH (B.), HELARD (J.F.), HALBERT (R.), LEMESLE (J.M.), Modulation et codage de canal pour la radiodiffusion sonore numérique vers les mobiles, *Colloque GRETSI'89*, Juan-Les-Pins, (12-16 June 1989).

[7] ETSI ETS 300 744, Digital video broadcasting (DVB): framing structure, channel coding and modulation for digital terrestrial television (DVB-T). (March 1997).

[8] ETSI, Project Broadband Radio Access Networks (BRAN); HIPERLAN Type 2, Technical specification; Physical layer, (Oct. 1999).

[9] IEEE, supplement to standard for telecommunications and information exchange between systems – LAN/MAN specific requirements – Part 11: wireless MAC and PHY specifications: high speed physical layer in the 5 GHz band, P802.11a/D7.0, (July 1999).

[10] CHOW (J.S.), TU (J.C.), CIOFFI (J.M.), A discrete multitone transceiver system for HDSL applications, *IEEE JSAC*, **9**, pp. 895-908, (Aug. 1991).

[11] YEE (N.), LINNARTZ (J.P.), FETTWEIS (G.), Multicarrier CDMA in indoor wireless radio networks, *Proceedings of IEEE PIMRC'93*, pp. 109-113, Yokohama, Japan, (1993).

[12] FAZEL (K.), PAPKE (L.), On the performance of convolutionnally-coded CDMA/OFDM for mobile communication system, *Proceedings of IEEE PIMRC'93*, pp. 468-472, Yokohama, Japan, (1993).

[13] FAZEL (K.), Performance of CCDMA/OFDM for mobile communication system, *Proceedings of IEEE ICUPC'93*, pp. 975-979, Ottawa, Canada, (1993).

[14] CHOULY (A.), BRAJAL (A.), JOURDAN (S.), Orthogonal multi-carrier techniques applied to direct sequence spread spectrum CDMA system, *Proceedings of IEEE Globecom'93*, pp. 1723-1728, Houston, USA, (1993).

[15] DASILVA (V.M.), SOUSA (E.S.), Performance of orthogonal CDMA codes for quasi-synchronous communication systems, *Proceedings of IEEE ICUPC'93*, pp. 995-999, Ottawa, Canada, (1993).

[16] VANDENDORPE (L.), Multitone direct sequence CDMA system in an indoor wireless environment, *Proceedings of the first IEEE symposium of communications and vehicular technology*, pp. 411-418, Delft, The Netherlands, (1993).

[17] HARA (S.), PRASAD (R.), Overview of multicarrier CDMA, *IEEE Communications Magazine*, pp. 126-133, (Dec. 1997).

[18] KAISER (S), OFDM-CDMA versus CDMA: performance evaluation for fading channels, *IEEE International Conference on Communications,* pp. 1722-1725 (June 1995).

[19] OH (J-Y.), LIM (M-S.), The bandwidth efficiency increasing method of multicarrier CDMA and its performance evaluation in comparison with DS-CDMA with Rake receivers, *Proceedings of IEEE Vehicular Technology Conference*, pp. 561-565, (1999).

[20] HELARD (J.-F.), BAUDAIS (J.-Y.), CITERNE (J.), Linear MMSE detection technique for MC-CDMA, *Electronics letters,* **36**, n° 7, pp. 665-666, (March 2000).

[21] YEE (N.), LINNARTZ (J.P.), Wiener filtering of multi-carrier CDMA in a Rayleigh fading channel, *Proceedings of IEEE PIMRC'94*, **4**, pp. 1344-1347, (Sept. 1994).

[22] LE GOUABLE (R.), HELARD (M.), Performance of MC-CDMA Systems in Multipath Indoor Environments – Comparison with COFDM-TDMA System, *3G Mobile Communication Technologies,* London, (March 2000).

[23] LE GOUABLE (R.), HELARD (M.), Performance of Single and Multi-User Detection Techniques for a MC-CDMA System over Channel Model used for HIPERLAN2, *Proceedings of ISSSTA2000, New-Jersey, USA,* (Sept. 2000).

[24] HELARD (J.-F.), BAUDAIS (J.-Y.), CITERNE (J.), Comparaisons des performances de différentes Techniques de détection Appliquées à un signal AMRS à Porteuses Multiples dans un Canal de Rayleigh", *GRETSI'99,* Vannes, France, (Sept. 1999).

[25] BAUDAIS (J.-Y.), HELARD (J.-F.), CITERNE (J.), An improved linear MMSE detection technique for Multi-Carrier CDMA systems: comparison and combination with interference cancellation, *European Transactions on Telecommunications,* Special issue on Multi-Carrier Spread-Spectrum, **11**, n° 6, pp. 547-554, (Dec. 2000).

[26] MOTTIER (D.), CASTELAIN (D.), A spreading sequence allocation procedure for MC-CDMA transmission systems, *Proceedings of the VTC 2000*, Boston, pp. 1270-1275, (Sept. 2000).

[27] KAISER (S), KRZYMIEN (W.A.), Performance effects of the uplink asynchronism in a spread spectrum multi-carrier multiple access system, *European Transactions on Telecommunications,* Special issue on Multi-Carrier Spread-Spectrum, **10**, n° 4, (July/Aug. 1999).

[28] DASILVA (V.), SOUSA (E.S.), Multicarrier orthogonal CDMA signals for quasi-synchronous communication systems, *IEEE Journal on Selected Areas in Communications*, **12**, pp. 842-852, (June 1994).

[29] SOUROUR (E.), NAKAGAWA (M.), Performance of orthogonal multicarrier CDMA in a multipath fading channel, *IEEE Trans. on Communications*, **44**, n° 3, pp. 356-367, (March 1996).

[30] Kondo (S.), Milstein (L.B.), Performance of multicarrier DS-CDMA systems, *IEEE Trans. on Communications*, **44**, n° 2, pp. 238-246, (Feb. 1996).

[31] Proakis (J.G.), Digital Communication, *McGraw Hill*, 1995.

[32] Cornic (A.C.), Grant (P.M.), Thomson (J.S.), A Novel Interference Cancellation Receiver for Uplink Multi-Carrier CDMA, *Proceeding of ISSSTA 2000*, Parsippany NJ USA, pp. 155-158, (Sept. 2000).

[33] Patel (P.) and Holtzman (J.), Performance comparison of a DS-CDMA system using a successive interference cancellation (IC) scheme and a parallel IC scheme under fading, *Proceedings of IEEE ICC'94*, New-Orleans, USA, (May 1994).

[34] Baudais (J.-Y.), Helard (J.-F.), Procédé d'égalisation dans des récepteurs utilisant une combinaison des techniques de modulations à porteuses multiples et à accès multiple par répartition de codes. French patent n° 99/11689, (Sept. 1999) and n° 99/15919, (Dec. 1999).

[35] Cost 207, Digital land mobile radio communications, tech. rep., Luxembourg: Office for Official Publications of the European communities.

[36] UMTS standard TR 101 112 V3.2.0, pp. 39-43, (April 1998).

[37] Maxey (J.J.), Ormondroyd (R.F.). Multi-carrier CDMA using convolutional coding and interference cancellation over fading channels. *First Workshop on Multi-Carrier Spread-Spectrum,* Oberpfaffenhofen, Germany, pp. 89-96, (April 1997).

[38] Kaiser (S.), Multi-Carrier CDMA Mobile Radio Systems – Analysis and Optimization of Detection, Decoding and Channel Estimation, PhD thesis, University of Munich, Germany, (1997).

[39] Berrou (C.), Glavieux (A.), Thitimajshima (P.), Near Shannon limit error-correcting coding and decoding: turbo codes, *ICC'93,* Geneva, (1993).

[40] Le Gouable (R.), Helard (M.), Performance of a MC-CDMA system including Turbo Codefor wireless indoor communications. *ECWT* Paris, (Sept. 2000).

Chapter 4

Interpretations and performances of linear reception in downlink TD-CDMA and multi-sensor extensions

L. Ros and G. Jourdain

Images and Signals Laboratory, Domaine University, France

M. Arndt

France Télécom R&D

I. Introduction

The normalization group 3GPP is preparing the standards for the third generation of radio-mobile system, UMTS (Universal Mobile Telecommunications System). Among them, the TDD (Time Division Duplex) mode [21] permits, for one or several users, the simultaneous transmission, during each TDMA slot, of spread QPSK symbols with different CDMA signatures. The uplink and downlink work at the same carrier frequency around 2 GHz, with quadrature modulation by two 3.84 Mchip/sec-binary streams of 5 MHz-spectral width. A synchronization channel and a midamble, transmitted at the slot rate, make the synchronization and channel estimation easier, independently of the symbol estimation mainly considered here. The spread spectrum systems were initially introduced in a single-user situation in order to ensure discretion or to fight against undesired channel effects [2]. They were calibrated, as far as possible, so that the symbol time was superior to the temporal spread of the channel, in such a way that the receiver was simply based on a matched filtering (MF).

Therefore, the transmission was robust but featured a very bad spectral efficiency.

In contrast, the systems which are considered today should provide an increase in the network global capacity. So, the spread messages from different users are superimposed[1], in a smaller number but as close as possible as to the spreading factor; this results in a non-negligible part of Multiple Access Interference (MAI). Moreover, the channel spread may go beyond the symbol time in the vehicular environment [20], leading to substantial Inter-Symbol Interference (ISI). To compensate interference and better detect each user's symbol, linear joint detection with temporal depth is a receiving strategy which uses the knowledge of the codes of other users, whereas the simple matched filter considers the MAI and ISI simply as wideband noise. The linear joint detection is not optimal for minimizing the Bit Error Rate (BER) but ensures a good compromise between performance and complexity, especially in TDD mode where the codes are short (16-chip long maximum). Nevertheless, interference cancelling, complete or partial, leads to an amplification of thermal additive noise. This effect can be largely reduced by using multisensor coherent processing on the mobile, which improves the waveform orthogonality and facilitates the system inversion. We only consider non-adaptive antenna for the base station transmitter.

The objectives of this chapter are:

(A) to interpret the CDMA system in downlink (in Section IV), usually with a frequency approach, natural in spread spectrum systems. We will establish the wide-band equivalent model, leading to an analogy with source separation in antenna processing, and the symbol-band equivalent model which is identical to the one not featuring spreading systems,

(B) to make a synthesis of the theoretical structures and performances of "classical" linear receivers for selective channels (in Sections III, V and VI), established from models described in (A). An exhaustive synthesis has been carried out by the authors of [4] but from block temporal formulation using a transfer matrix of all the symbols in the slot. The frequency approach permits here simplification of the model and to take into account the selective channels naturally,

(C) to evaluate the average performances in Vehicular and Indoor environment models (in Section VII) by application of the expressions of (B) in order to measure the interest of joint detection and multi-sensor reception over selective and non-selective channels.

II. Multi-user transmission model

Continuous-time baseband representation (complex envelope) of the received signal, observed during one time slot (containing M symbols), is modelled as:

1. To simplify the description we abusively associate, in the whole article, one active code to one user. In real applications, several codes may be affected to the same user in order to increase its information rate (multi-code situation).

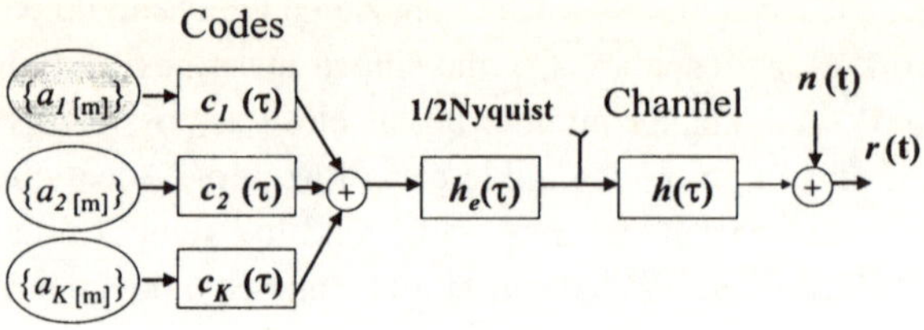

Figure 1. Baseband model of transmission.

(1) $r(t) = Ts \cdot \sum_{k=1}^{K} \sum_{m=1}^{M} a_{k[m]} \cdot g_k(t - mTs) + n(t)$

where

– $a_{k[.]}$ are the QPSK symbols transmitted by the *k-th* source at the "symbol times" mTs,

– $g_k = c_k * h_e * h$ is the overall *wide-band waveform of the k-th user*. It results from the convolution between the *k-th* spreading code, the half Nyquist (square-root raised cosine filter "RRC" with roll-off r_{off}), and the channel,

– $n(t)$ is a baseband additive complex noise.

We consider the following context of downlink multi-user communication:

– the number of users (or active codes) K is less or equal to the spreading factor Q;

– the K active codes (taken among a set of Q known codes) are supposed known at the receiver, which is not necessarily the common situation (necessity of a control channel);

– the codes of different users are made from Q binary complex entity named "chips". The codes are linearly independent and orthogonal at zero delay (algebraically meaning), which is expressed by:

(2) $\gamma_{ik[0]}^{code} = \sum_{q=0}^{Q-1} c_{i[q]} c_{k[q]}^* = \delta_{ik}$

where δ is the Kronecker symbol whereas δ(.) is the Dirac distribution. The impulse response of one code lasts for Ts, and is defined for the *k-th* user by:

$c_k(\tau) = \sum_{q=0}^{Q-1} c_{k[q]} \delta(\tau - qTc)$, where $Tc = Ts/Q$ is the chip duration;

– the transmitted symbols are stationary, with zero mean and power A^2; they are uncorrelated temporally and from one user to another, and also uncorrelated with the additive noise;

– "downlink" with "non-intelligent Base Station": all K user's signals appear at each mobile sensor after transmission over the same propagation channel and with the same power. The channel h is supposed to be deterministic, time-invariant during the observation (1 slot), linear and causal with a temporal spread limited to W_c chips. The structures studied at the sequel need only these assumptions for the channel. Nevertheless in radio-mobile communications, the channel is generally constituted by multiple paths, and its impulse response is modelled by:

(3) $$h(\tau) = \sum_{i=1}^{L_t} \alpha_i \cdot \delta(\tau - \tau_i)$$

where L_t is the number of paths, α_i are the complex amplitudes and τ_i the propagation delays of different paths, with $\tau_i \leq W_c \cdot Tc$, $\forall i \in \{1, \ldots, L_t\}$;

– the complex noise is an additive white Gaussian noise, with two-sided power spectral density $2N_0$ (corresponding to a two-sided psd $\frac{N_0}{2}$ for the real noise on band carried at the entrance of the receiver);

– two slots of symbols are isolated by a guard time above $W_c \cdot Tc$;

– we suppose perfect timing recovery (using the synchronization channel) and also perfect channel estimation (using the midamble).

We suppose by convention only one desired code (or desired user), the code number one.

We define the *average energy per bit* at the entrance of the receiver (on band carried) for the desired user:

$E_b = \frac{1}{4} A^2 Ts\ \gamma_{11[0]}$, where $\frac{\gamma_{11[0]}}{Ts}$ is the zero-delay auto-correlation of the waveform $g_1(\tau)$ of user "1".

Note: the assumption of no variation of the channel during the slot ($T_{slot} = 667\ \mu sec$) is fully justified for a mobile moving with a speed υ_m of 3 km/h. Indeed, for a carrier radio-wave (celerity υ_0) around f_0 = 2 GHz, the Doppler spread is defined by $\Delta f_d = \frac{\upsilon_m}{\upsilon_0} \cdot f_0$ and the channel coherence time $T_{coh} = \Delta f_d^{-1}$ is around 200 msec. The hypothesis limit occurs with vehicular moving at υ_m = 120 km/h ($T_{coh} \approx 5$ msec) since the complex amplitudes of the paths may suffer from phase rotation up to

$\frac{\frac{1}{2} T_{slot}}{T_{coh}} \cdot 360°$ around ±25° during one 1/2 slot, in the case of a deterministic Doppler model. We might need to update the knowledge of the complex amplitudes several times in the slot for optimizing the symbol detection.

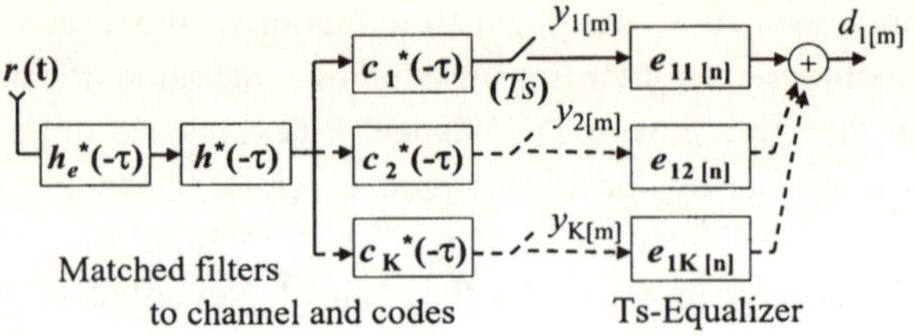

***Figure 2**. Joint linear detector (with $\gamma_{11[0]} = 1$).*

On the other hand, since the mobile speed υ_m is always negligible compared to the wave celerity υ_0, the variation of the path delay $\frac{\upsilon_m}{\upsilon_0} . T_{slot}$ is negligible during the slot (around $10^{-4}Tc$).

III. Structure of the joint linear detection

With infinite response, the symbol by symbol linear detector [17] can take the shape of an imposed structure with one matched filter bank followed by a bank of discrete equalizing filters working at symbol rate, as seen in Figure 2. It is easy to verify this, under classical criteria (Zero-Forcing, MMSE).

III.1. Channel- and code-matched filter bank

The receiver front-end applies to $r(t)$ a bank of continuous filters g_k^H, $k=1,..., K$ matched to each global waveform, made by the chip pulse (RRC), the channel response and each of the K codes. By convention, the exponent $(.)^H$ represents hermitian transform for a function h (i.e. $h^H(\tau) = h^*(-\tau)$: matched filter) just as the hermitian transposition for vectors or matrix. The K outputs of filters g_k^H sampled synchronously at symbol time, constitute a sufficient statistic of the continuous signal received $r(t)$ for the symbol estimation [1, 16]:

(4) $$y_{k[m]} = (g_k^H * r)(t)\big|_{t=mTs}$$

Thus, knowledge of $\{y_{k[m]}\}$, $k=1,\ldots,K$, $m=1,\ldots,M_+$ is sufficient to know the likelihood function of the received signal and to carry out the optimal processing which minimizes the error probability (or BER). For the treatment of one slot, we must consider a number M_+ of samples slightly greater than the effective number M of symbols because of the channel temporal spread and because of the matched filter. In this chapter, we will not consider the optimal theoretical post-processing, i.e. the Maximum Likelihood Sequence Estimator, but only the detectors which treat linearly $\{y_{k[m]}\}$ and make the decisions symbol by symbol.

The *Ts*-sampled outputs $\{y_{k[m]}\}$ are expressed from discrete time convolutions of the transmitted symbols, disturbed by an additive noise:

$$(5) \qquad y_{k[m]} = \sum_{i=1}^{K} (\gamma_{ik} * a_i)_{[m]} + \eta_{k\,[m]}$$

where $\gamma_{ik\,[n]}$ is the *sampling, at symbol time, of the continnous and deterministic cross-correlation function* $\gamma_{ik}^{cont}(\tau) = (g_i * g_k^H)(\tau)$ between the wide-band waveforms of users *i* and *k*:

$$(6) \qquad \gamma_{ik\,[n]} = Ts \cdot \gamma_{ik}^{cont}(\tau)_{|\tau = nTs}$$

After matched filtering and synchronous sampling, equation (5) shows that all the system can be modelled at discrete time *Ts*. The series $\{\gamma_{ik\,[n]}\}$, $\forall n \in \mathbb{Z}$ providing the discrete filtering, has no dimension. Through a multi-path channel, the waveforms are generally not orthogonal and some ISI and MAI interference occurs with secondary terms γ_{11} at non-zero delays and cross-channel terms γ_{1i}, $i \neq 1$ at the different delays respectively:

$$y_{1[m]} = \underbrace{\gamma_{11[0]} \cdot a_{1[m]}}_{desired\ part} + \underbrace{\sum_{n \neq m} \gamma_{11[m-n]} \cdot a_{1[n]}}_{ISI} +$$

$$(7) \qquad \underbrace{\sum_{i=2}^{K} \sum_{n} \gamma_{i1[m-n]} \cdot a_{i[n]}}_{MAI} + \underbrace{\eta_{1[m]}}_{Noise}$$

The necessity and the difficulty of the equalization will be entirely conditioned by the values $\{\gamma_{ik[n]}\}$.

Recalling that:

* In the case of a single-user system or without joint detection, only the first branch is used, operating a coherent combination relative to the different paths, called Rake; $y_{1[m]}$ may then be the decision variable.

* In the case of a multi-user system but a single-path channel, the duration of the code limited to *Ts* (and respect of the Nyquist criterion at the chip time for h_e) leads to no interference at non-zero delay since $\gamma_{ik[n]} = 0$, $\forall n \neq 0$. Moreover, when the codes are orthogonal, there is no interference at zero delay since $\gamma_{ik[0]} = \delta_{ik}$: only the first branch of the MF bank is useful, reduced to the receiving filter RRC followed by correlation with the desired code.

* In the case of a multi-path channel with weak temporal spread relative to symbol time, the contribution at non-zero symbol time delays are negligible and only the MAI at the zero delay may be considered. It is similar to a situation where the waveforms would last *Ts*, but without being orthogonal: this is the case studied by most authors [16, 6, 5], corresponding to a well calibrated system, but it leaves out a number of situations in a vehicular environment with the TDD-UMTS parameters.

III.2. Bank of discrete equalizers (Ts)

In order to form a decision variable $d_{1[m]}$ used to estimate the user's one QPSK symbols $a_{1[m]}$ by quadrant detection, an equalizer linearly combines the outputs $\{y_{k[m]}\}$ of the K branches of the matched filter-sampler bank, after passing them through a bank of discrete filters $\{e_{1k[n]}\}$ working at symbol time:

(8) $$d_{1[m]} = \frac{1}{\gamma_{11[0]}} \cdot \sum_{k=1}^{K} (e_{1k} * y_k)_{[m]}$$

Having described the structure of the detector, we will now derive and interpret equivalent models of the CDMA system before setting the performances and expressions of the multi-user equalizer.

IV. Interpretations and models of CDMA

During transmission, equation (1) expresses the multiple access based on orthogonal waveforms (symbol pulse-shapes) from one user to another, for an ideal channel. For the particular CDMA case, each pulse g_k, for $k = 1$ to K, occupies all the available time-frequency plane($Ts \cdot \frac{Q}{Ts}$) instead of being located in time (width $\frac{Ts}{K}$ in TDMA) or in frequency (width $\frac{1}{Ts}$ in FDMA). The number ($\frac{Q-K}{Q}$) plays the role of a code margin with regard to system saturation, same as the margins provided by the guard times or the guard bands in TDMA or FDMA.

By introducing in (10) the impulsive form $a_k(t)$ of the different user symbols, the signal received defined in (1) can be expressed classically as a time continuous convolution between the symbols and the global wide-band waveforms, disturbed by an additive noise:

(9) $$r(t) = \sum_{k=1}^{K} (g_k * a_k)(t) + n(t)$$

with

$$(10) \qquad a_k(t) = Ts \cdot \sum_{m=1}^{M} a_{k[m]} \cdot \delta(t - mTs)$$

This expression allows describtion of the chain as a cascade of linear filters excited by the impulsive symbols streams, exactly the same way as a linear modulation without spreading. The difference here is that the pulse shape is wide-band compared to the symbol band of width $\frac{1}{Ts}$. This formulation will facilitate the frequency transformation in the following paragraph.

IV.1. Interpretation at transmission (multi-band transmission)

The result of a single observation during the slot can be analysed in the frequency domain with the Fourier transform. The signals $a(t)$ and $n(t)$ are considered here as deterministic (being particular occurences during a slot of random variables), which avoids treatment with (cyclic) autocorrelation functions of the received signal. The Fourier Transform of equation (9) gives:

$$(11) \qquad \hat{r}(f_w) = \sum_{k=1}^{K} \hat{g}_k(f_w) \cdot \hat{a}_k(f_w) + \hat{n}(f_w)$$

where by convention:

- f_w is used for any frequency while f is a frequency of the symbol-band, included in $[-\frac{1}{2Ts}, +\frac{1}{2Ts}]$. So, for all f_w there is a single frequency f and a single relative integer q such that $f_w = f + \frac{q}{Ts}$,
- the continuous Fourier transform of a function h in continuous time or discrete time (then obtained via Z transform) is noted $\hat{h}(.)$.

The transmission bandwidth, defined by the support of $\hat{g}_k(f_w)$ is limited to the band of the half Nyquist filter, that is $Q_r \cdot \frac{1}{Ts}$ where Q_r is the integer part of $(1 + r_{off})$. Q, expressing the filter bandwidth excess. With the UMTS parameters $Q = 16$ and $r_{off} = 0.22$, Q_r is odd and equal to 19.

The Q_r symbol bands (of width $\frac{1}{Ts}$) which comprise the wide-band are then centred around the frequencies $f_w = -\frac{Q_{r/2}}{Ts}, \ldots, 0, \ldots, +\frac{Q_{r/2}}{Ts}$ with the definition of $Q_{r/2}$ such that $Q_r = 2Q_{r/2} + 1$.

Note: Taking into account the RRC attenuation, the portions not included in this Q_r bands are ignored. If one does not want to lose any information, we might have taken $Q_r + 1 = 20$ instead of Q_r and choose for f a range equal to $[0, +\frac{1}{Ts}]$.

Due to the characteristic of "sampling" (cf (10)) of $a_k(t)$ at the symbol rate (or the cyclostationarity of the digital communication signals), $\hat{a}_k(f_w)$ is periodic with a period of $\frac{1}{Ts}$:

(12) $$\hat{a}_k(f) = \hat{a}_k(f + \frac{q}{Ts}),\ \forall q \in \mathbb{Z}$$

Expression (11) is then written inside the useful wide-band:

(13) $$\hat{r}(f + \frac{q}{Ts}) = \sum_{k=1}^{K} \hat{r}_k(f + \frac{q}{Ts}) + \hat{n}(f + \frac{q}{Ts})$$

with $\forall q = -Q_{r/2}, \ldots, +Q_{r/2}$:

(14) $$\hat{r}_k(f + \frac{q}{Ts}) = \hat{g}_k(f + \frac{q}{Ts}) \cdot \hat{a}_k(f)$$

Equation (14) expresses the *frequency diversity of order* Q_r inherent to spread spectrum systems: for a given user "k", *the same binary message* $a_{k[m]}$, $m = 1, \ldots, M$, with Fourier transform $\hat{a}_k(f)$ on the complete symbol band, *is transmitted in* Q_r *different symbol-bands*. Generally (with weak excess bandwidth or roll-off), the portions outside the chipband (width $\frac{1}{Tc}$) are attenuated and only a diversity of order Q is considered, according to the temporal structure of codes, measured at the chip rate.

In an equivalent way, emphasis can be put on diversity by means of a *multi-carrier modulation* concept. From the periodicity of $\hat{a}_k(f)$ and the impulsive form of the code $c_k(\tau)$ with finite length Ts, the k-*th* user's signal, before RRC and channel filtering, is expressed in frequency domain by:

(15) $$\hat{e}_k(f_w) = (\hat{c}_k^{per} * \hat{a}_k^{rect})(f_w)$$

$$\text{with: } \hat{c}_k^{per}(f_w) = \sum_n \sum_{q=-\frac{Q}{2}}^{+\frac{Q}{2}-1} \hat{c}_k(\frac{q}{Ts}) \cdot \delta(f_w - \frac{q}{Ts} - \frac{n}{Tc})$$

$$\hat{a}_k^{rect}(f_w) = \hat{a}_k(f_w) \cdot \text{sinc}(\pi Ts\, f_w)$$

Unlike OFDM (Orthogonal Frequency Division Multiplexing), for a given user, the sub-carriers are modulated by the same information (complex binary message). There are spaced by $\frac{1}{Ts}$ and have complex amplitudes $\hat{c}_k(\frac{q}{Ts})$ given by the DFT (Discrete Fourier Transform) of the Q chips of the code. The messages of different users will easily be separated at the receiver, at least over ideal channel, thanks to the codes orthogonality (independency) property, which is expressed in the frequency-domain by the linear orthogonality (independency) of the different sets of Q sub-carriers $\{\hat{c}_k(\frac{q}{Ts})\}$.

Note: we can establish equation (15) more naturally by representing the spread-spectrum as the result of the multiplication of the continuous symbol stream $a_k^{rect}(t)$ (shaped into a rectangular pulse of length *Ts*) by the *periodic* version of the code $c_k^{per}(t)$ (with period *Ts*). The expression of the periodic code in Fourier series gives then the multi-carrier modulation scheme directly. This other equivalent representation of the spreading system, i.e. the "multiplicative representation", is closer to the design, and probably more traditional [3]. However, it is more difficult to manipulate theoretically than the representation with linear filters, generally adopted.

IV.2. Interpretation at reception

Assuming a perfect knowledge of the channel and timing synchronization, the outputs of the bank of MF-samplers can be synthesized from a linear transformation of the transmitted symbols, without mentioning the wide-band received signal. The temporal expression, at discrete time *Ts*, of the "symbol-band model" has been established in (5).

The transfer $\hat{\gamma}_{ik}(f)$ between the symbols of the source "*i*" and the output of the branch "*k*" is expressed in frequency-domain by Fourier transform of equation (6) as the *aliasing of the wide-band cross-spectrum* $\hat{\gamma}_{ik}^{cont}(f_w)$ *in the symbol-band*:

$$(16) \qquad \hat{\gamma}_{ik}(f) = \sum_{q=-Q_{r/2}}^{+Q_{r/2}} \hat{\gamma}_{ik}^{cont}\left(f - \frac{q}{Ts}\right)$$

$$(17) \qquad \text{with:} \qquad \hat{\gamma}_{ik}^{cont}(f_w) = \hat{g}_i\, \hat{g}_k^*(f_w)$$

$$= |\hat{h}|^2 |\hat{h}_e|^2\, \hat{c}_i\, \hat{c}_k^*(f_w)$$

We can now explain the notions of aliased equivalent channels in the particular case of CDMA.

IV.2.1. Aliased Tx/Rx global channels

At the first branch output, the transfer $\hat{\gamma}_{11}(f)$ is equivalent to an aliased Trans-Receiving Global Channel (aT/RGC) in the symbol band for the desired user, the transfer $\hat{\gamma}_{i1}(f)$ to an aT/RGC for the interfering user "$i \neq 1$". Expressed in frequency, the absence of ISI on the branch "1" corresponds to an aT/RGC white for the desired user, and the absence of MAI corresponds to the nullity of the aT/RGC for all the interfering users. These properties are absolutely false for the T/RGC before aliasing, $\hat{\gamma}_{ik}^{cont}(f_w)$, because of inter-chip and inter sub-chip interference on the wide-band channels outputs before sampling. It simply expresses, for the desired user, a coherent recombination of the sub-carriers and

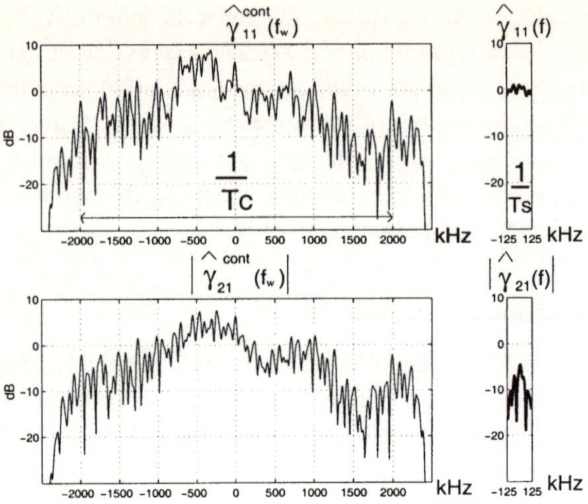

Figure 3. *Tx/Rx Global Channel example in* VB.

for the interfering users, a destructive re-combining. With a single-path channel, the orthogonality (2) and the finite length of the codes ensures these conditions. It is easy to show using the Poisson formula that $\forall f \in [-\frac{1}{2Ts}, +\frac{1}{2Ts}]$:

$$(18)\quad \sum_{q=-\frac{Q}{2}}^{+\frac{Q}{2}-1} (\hat{c}_k \cdot \hat{c}_j^*)\left(f - \frac{q}{Ts}\right) = \gamma_{kj\,[0]}^{code} = \delta_{kj}$$

Respect of the first Nyquist criterion, relative to the chip time Tc for $h_e * h_e^H$, leads then to:

$$(19)\quad \sum_{q=-Q_{r/2}}^{+Q_{r/2}} (\hat{c}_k \cdot |\hat{h}_e|^2 \cdot \hat{c}_j^*)\left(f - \frac{q}{Ts}\right) = \delta_{kj}$$

For the desired user, equation (19) simply expresses the first Nyquist criterion relative to the symbol time Ts for $g_1 * g_1^H$ *with an ideal channel.* When the channel is not single-path, the aliasing equation (16) explains the robustness of the spread-spectrum to selective channels: the aT/RGC seen by the desired user symbols is finally a channel with width $\frac{1}{Ts}$ obtained from the coherent superposition of Q_r channels with equal width, averaging the selectivity imposed to the different bands.

Figure 3 shows the T/RGC and aT/RGC on branch one for the desired user and for the interferer "2", with a channel taken from the Vehicular B model, characterized by a temporal spreading around 5 *Ts*. The power of Interference and Noise (IN) at the first branch output is derived from the temporal expression of the ISI, MAI and Noise terms defined in (7):

$$(20) \qquad (IN) \triangleq E\{|y_{1[m]} - \gamma_{11[0]} \cdot a_{1[m]}|^2\}$$

$$= A^2 \sum_{n \neq 0} |\gamma_{11[n]}|^2 + A^2 \sum_{k=2}^{K} \sum_{n} |\gamma_{k1[n]}|^2 + 2N_0 \cdot \frac{\gamma_{11[0]}}{Ts}$$

The expression of (IN) VS frequency, established with the Parseval relation applied to equation (20), shows that the Interferference and Noise power depends only on the form of the aT/RGC:

$$(IN) = A^2 Ts \int_{\frac{1}{Ts}} |\hat{\gamma}_{11}(f) - \gamma_{11[0]}|^2 df$$

$$+ A^2 Ts \sum_{k=2}^{K} \int_{\frac{1}{Ts}} |\hat{\gamma}_{k1}(f)|^2 df + 2N_0 \cdot \int_{\frac{1}{Ts}} \hat{\gamma}_{11}(f) df$$

The power of ISI is derived from the quadratic frequency area between the desired aT/RGC and an ideal flat (white) one: it results thus from the non-whiteness of the desired aT/RGC. The power of MAI is derived from the sum of the quadratic frequency area of interfering aT/RGC relative to the branch " 1". The noise power is derived from the Noise Equivalent Bandwidth: $\gamma_{11[0]} \cdot \frac{1}{Ts}$ of the branch "1", i.e. from the frequency area in the symbol band of the desired aT/RGC.

IV.3. Equivalent matrix models

For compacting and simplifying the representation of the downlink CDMA system, we can use two equivalent matrix models, named wide-band model and symbol-band model. The diversity can well be represented by the wide-band model, by showing the Q_r bands containing the signal received. It may be useful when some parameters (channel, codes, timing synchronization) are not identified. Otherwise, being the context of this article, the representation model in the symbol band will be sufficient to establish the expressions and performances of the equalizers.

IV.3.1. Wide-band model

The Q_r components of the signal received at frequency f in the different bands (equations 13 and 14) are expressed algebraically by:

(21) $$\underline{\hat{r}}(f) = \underline{\underline{G}}(f)\,\underline{\hat{a}}(f) + \underline{\hat{n}}(f)$$

where

$$\underline{\underline{G}}(f) \triangleq [\underline{\hat{g}}_1(f), \ldots, \underline{\hat{g}}_K(f)],$$

$$\underline{\hat{g}}_k(f) = [\hat{g}_k(f - \frac{Q_{r/2}}{Ts}), \hat{g}_k('''' + \frac{1}{Ts}) \ldots, \hat{g}_k(f + \frac{Q_{r/2}}{Ts})]^{\mathrm{T}},$$

$$\underline{\hat{r}}(f) = [\hat{r}(f - \frac{Q_{r/2}}{Ts}), \ldots, \hat{r}(f + \frac{Q_{r/2}}{Ts})]^{\mathrm{T}},$$

$$\underline{\hat{n}}(f) = [\hat{n}(f - \frac{Q_{r/2}}{Ts}), \ldots, \hat{n}(f + \frac{Q_{r/2}}{Ts})]^{\mathrm{T}},$$

$$\underline{\hat{a}}(f) = [\hat{a}_1(f), \ldots, \hat{a}_K(f)]^{\mathrm{T}}.$$

The matrix $\underline{\underline{G}}(f)$ of size $Q_r \times K$ contains the wide-band waveforms relative to f in the different symbol-bands, and for the different users. It plays the role of a transfer matrix from the K sources towards the Q_r bands, for a given frequency of the symbol-band.

An analogy can be made with the source separation from the reception on an array of Q_r sensors (in place of Q_r frequency bands). In antenna processing domain [19, 8], each of the Q_r sensors receives a mixture of K sources but the latter arrive from different directional (spatial) vectors. In CDMA, each of the Q_r symbol-bands of the reception contains a mixture of the K sources (each band containing all the information for a given source) but the sources "arrive" from different "frequency vectors", imposed by the codes and the channel.

We find again some features which are also classical assumptions in antenna processing:

– the noise is uncorrelated from one band to another:

$$\frac{1}{T_{slot}} \cdot E\{\underline{\hat{n}}(f)\,\underline{\hat{n}}^H(f)\} = 2N_0 \cdot \underline{\underline{I}}_{Qr}$$

– the uncorrelation between the sources leads to:

$$\frac{1}{T_{slot}} \cdot E\{\underline{\hat{a}}(f)\,\underline{\hat{a}}^H(f)\} = (A^2 Ts) \cdot \underline{\underline{I}}_K$$

where $\underline{\underline{I}}_N$ denotes the identity matrix of size $N \times N$.

The beamforming favours one spatial direction for extracting the source "1" from the coherent recombining of the outputs of the Q_r sensors. In the same way in CDMA, the matched filter favours one "frequency signature" by coherently (in branch "1" for the desired user) re-combining the components of the Q_r bands or sub-carriers, which appears in (4) and can be formulated as:

$$(22) \quad \underline{\hat{y}}(f) = \underline{\underline{G}}^H(f)\,\underline{\hat{r}}(f)$$
$$= \underline{\underline{G}}^H(f)\,\underline{\underline{G}}(f)\,\underline{\hat{a}}(f) + \underline{\underline{G}}^H(f)\,\underline{\hat{n}}(f)$$

where

$$\underline{\hat{y}}(f) = [\hat{y}_1(f), \ldots, \hat{y}_K(f)]^T.$$

The equivalent diagram of transmission and reception for the wide-band model is given in Figure 4:

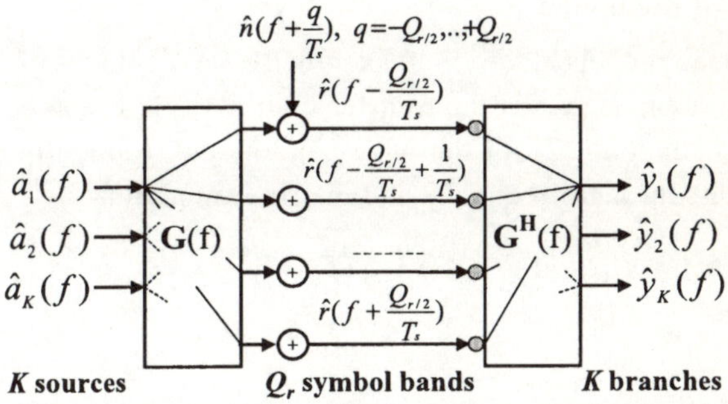

Figure 4. Transmission multi-band model.

IV.3.2. Symbol-band model

Spectral matrix of the aliased cross-channels

We define, for each frequency f belonging to [– 1/2 Ts; + 1/2 Ts], the normalized spectral matrix of the cross-channels aliased in the symbol-band. This matrix groups all the AT/RGC in the following form:

$$\underline{\underline{\Gamma}}(f) \triangleq \frac{1}{\gamma_{11[0]}} \cdot \begin{pmatrix} \hat{\gamma}_{11}(f) & \hat{\gamma}_{21}(f) & \cdots & \hat{\gamma}_{K1}(f) \\ \hat{\gamma}_{12}(f) & \hat{\gamma}_{22}(f) & \cdots & \hat{\gamma}_{K2}(f) \\ \hat{\gamma}_{1K}(f) & \hat{\gamma}_{2K}(f) & \cdots & \hat{\gamma}_{KK}(f) \end{pmatrix}$$

This matrix is of a $K \times K$ size, hermitian and non-negative definite. It will be with full rank and so positive definite for all f if the spectrum $|\hat{h}(f_w)|^2$ of the propagation channel does not have zero in the wide-band. Note that the absence of zero is a sufficient but not a necessary condition. On the contrary, if the spectrum of the propagation channel has a common zero at a same frequency f in more than $(Q_r - K)$ different symbol-bands, the matrix is necessarily with rank deficient for this specific frequency. Therefore, except for pathological channels, the *normal* rank (*i.e.* for almost all f) will be full, equal to K.

In terms of interference for the whole users, ISI corresponds to the "non-whiteness" of the diagonal terms of $\underline{\underline{\Gamma}}(f)$, and MAI to the non-zero non-diagonal terms.

The formulation of $\underline{\underline{\Gamma}}(f)$ from $\underline{\underline{G}}(f)$ is expressed in a matrix form from the aliasing relation (16):

$$\underline{\underline{\Gamma}}(f) = \frac{1}{\gamma_{11[0]}} \cdot \underline{\underline{G}}^H(f)\,\underline{\underline{G}}(f)$$

Matrix form of the model

The symbol-band model (Figure 5) links directly the K sources of symbols to the outputs of the K branches via the transfer matrix $\underline{\underline{\Gamma}}(f)$. It can be derived from Figure 4 (i.e. equation (22)) of the wide-band model or directly from equation (5) after passage through frequency and matrix formulation:

(23) $$\underline{\hat{y}}(f) = \gamma_{11[0]} \cdot \underline{\underline{\Gamma}}(f)\,\underline{\hat{a}}(f) + \underline{\hat{\eta}}(f)$$

where

$$\underline{\hat{\eta}}(f) = [\hat{\eta}_1(f), \ldots, \hat{\eta}_K(f)]^T = \underline{\underline{G}}^H(f)\,\underline{\hat{n}}(f).$$

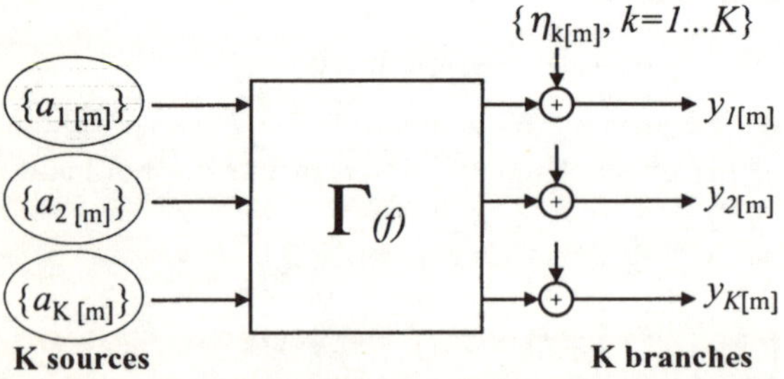

***Figure 5**. Symbol-band model of the transmission.*

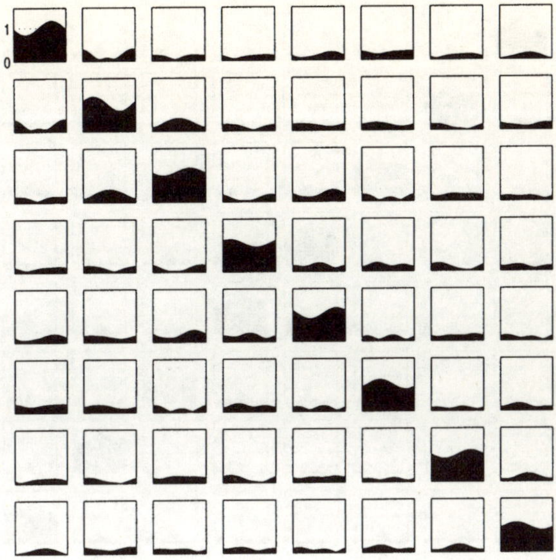

Figure 6. *Matrix* $|\underline{\underline{\Gamma}}(f)|$ *example in Pedestrian B.*

The equivalent discrete noise $\eta_{k[m]}$, artificially reported in addition on each branch ouptut is gaussian, correlated temporally and from one branch to another. Its cross-spectral density, from branch "*i*" to branch "*k*" is

$$\frac{E\{\hat{\eta}_i(f)\cdot\hat{\eta}_k^*(f)\}}{T_{slot}} = 2N_0\cdot\hat{\gamma}_{ik}(f), \text{ is under matrix form:}$$

$$\frac{1}{T_{slot}}\cdot E\{\hat{\underline{\eta}}(f)\,\hat{\underline{\eta}}^H(f)\} = 2N_0\cdot\gamma_{11[0]}\cdot\underline{\underline{\Gamma}}(f)$$

Two examples of the matrix $|\underline{\underline{\Gamma}}(f)|$ are given for $K = 8$ users (with linear scales): in Figure 6, the temporal spreading of the channel (one realization of the Pedestrian B channel according to the ETSI model) is around 1 *Ts*.

In Figure 7, the temporal spreading is around 5 *Ts* (same Vehicular B channel as used in Figure 3). In the two examples, the interference is around the same order but we can verify that the coherence bandwidth in Vehicular B is about 5 times smaller.

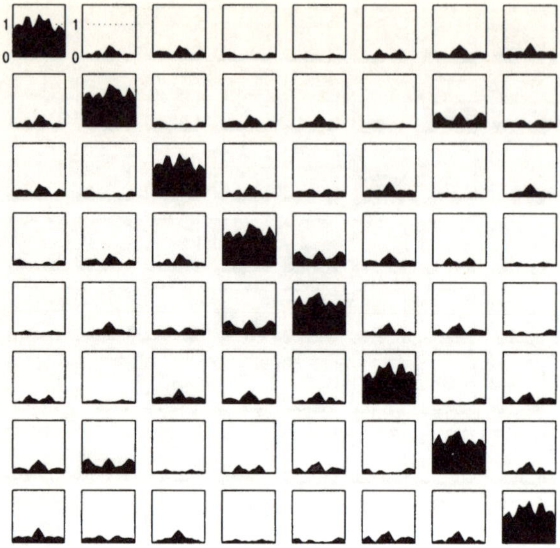

***Figure** 7. Matrix $|\underline{\underline{\Gamma}}(f)|$ example in Vehicular B.*

V. Expressions and performances of linear detections

V.1. Matched filter only

V.1.1 Signal to interference and noise ratio

The matched filter has the feature of maximizing the Signal to Noise Ratio (SNR) at *mTs* times, without considering the interference (ISI and MAI). To take the interference into account, the quality criterion of the Signal to Interference and Noise Ratio (SINR) is best appropriate.

For the desired user ("1"), the SINR in output of the first branch of the matched filter bank is expressed in the matrix $\underline{\underline{\Gamma}}(f)$ by:

$$(24) \qquad (SINR)_{MF} = \frac{A^2 \gamma^2_{11[0]}}{(IN)}$$

$$= (\frac{2E_b}{N_0}) \cdot \frac{1}{1 + (\frac{2E_b}{N_0}) \underline{\underline{\Sigma}}^T Ts \int_{\frac{1}{Ts}} |\underline{\underline{\Gamma}}(f)\underline{1} - \underline{1}|^2 df}$$

where $\underline{\sum}^T \triangleq [1, \ldots, 1]$, $\underline{1}^T \triangleq [1, 0, \ldots, 0]$

In the absence of ISI and MAI, the matched filter obtains the best SINR equal to $(\frac{2E_b}{N_0})$ as well as the ideal elementary Binary Error Probability.

V.2. Equalizers: ZF and MMSE

In the presence of interference, a numerical linear equalizer bank running at Ts completes the head of the base-band receiver, as set forth in Section III.2.

The decision variable $d_{1[m]}$ defined in (8) can be written as a global convolutive discrete transfer $\{glo_{1i[n]}, i = 1, \ldots, K\}$ for the symbols, disturbed by an additive discrete noise b_1:

(25) $$d_{1[m]} = \sum_{i=1}^{K} (glo_{1i} * a_i)_{[m]} + b_{1[m]}$$

with $glo_{1i[n]} \triangleq \frac{1}{\gamma_{11[0]}} \sum_{k=1}^{K} (e_{1k} * \gamma_{ik})_{[n]}$

The Mean Squared Error (MSE) is generally used to describe the performances of the equalizer. This energetic measure benefits from a simple quadratic formulation and presents simple solutions of minimization problems. The MSE in the output of the equalizer for the desired user is expressed in frequency terms by:

$$(MSE)_{Eq} \triangleq E\{|d_{1[m]} - a_{1[m]}|^2\}$$

$$= A^2 Ts \int_{\frac{1}{Ts}} |\hat{glo}_{11}(f) - 1|^2 df + A^2 Ts \sum_{k=2}^{K} \int_{\frac{1}{Ts}} |\hat{glo}_{1k}(f)|^2 df$$

$$+ 2N_0 \cdot \int_{\frac{1}{Ts}} |\hat{t}_1(f)|^2 df$$

The integration term in this expression is the power spectral density of the error (psd), comprised of a ISI term, a MAI term, and a noise term.

From matrix formulation:

– the vector $\hat{e}_{\underline{1}}{}^T(f) = [\hat{e}_{11}(f), \ldots, \hat{e}_{1K}(f)]$ is created from the transfer functions ofthe equalizing filters on the K branches,

– the global transfer function for the symbols of the user k is: $\hat{glo}_{1k}(f) = \hat{e}_{\underline{1}}{}^T(f) \cdot \underline{\underline{\Gamma}}(f) \cdot \underline{k}$

where $\underline{k}^T = [\underbrace{\ldots, 0, 1, 0, \ldots}_{k}]$

– the global transfer function $\hat{t}_1(f)$ for the noise (with periodicity $\frac{1}{Ts}$ due to the sampling) is such that:

$$|\hat{t}_1(f)|^2 = \frac{1}{\gamma_{11[0]}} \underline{\hat{e}_1}^H(f) \cdot \underline{\underline{\Gamma}}^*(f) \cdot \underline{\hat{e}_1}(f).$$

The psd of $b_{1[m]}$ is then $2N_0 \cdot |\hat{t}_1(f)|^2$.

From these expressions, it is easy to establish, by omitting (f) in the second member for the sake of clarity, that:

$$\text{(26)} \qquad \frac{psd}{A^2 Ts}(f) = \underline{\hat{e}_1}^H (\underline{\underline{\Gamma}}^* \underline{\underline{\Gamma}}^* + \frac{N_0}{2E_b} \cdot \underline{\underline{\Gamma}}^*)\, \underline{\hat{e}_1} - 2Re\{\underline{\hat{e}_1}^H \underline{\underline{\Gamma}}^* \underline{1}\} + 1$$

Notes:

* Interference power plus noise can be obtained simply from the MSE by subtracting the quadratic error term at zero delay: $(IN)_{Eq} = (MSE)_{Eq} - A^2|1-glo_{11[0]}|^2$.

* The global linear detector l_1 for the detection of the symbols of user " 1", including the MF-sampler bank and the equalizing filter bank, has a transfer function $\hat{l}_1(f_w)$ defined for the wide-band. It achieves actually, for one frequency f of the symbol-band, a linear combination of the Q_r sub-bands. Its expression under vector form of size Q_r (grouping $\hat{l}_1(f - \frac{Q_{r/2}}{Ts}), \ldots\ \hat{l}_1(f + \frac{Q_{r/2}}{Ts})$, is:

$$\text{(27)} \qquad \underline{\hat{l}_1}^T(f) = \frac{1}{\gamma_{11[0]}} \cdot \underline{\hat{e}_1}^T(f)\, \underline{\underline{G}}^H(f)$$

V.2.1. ZF equalizer

The optimal ZF equalizer minimizes the MSE, under the constraint of cancelling the ISI and MAI. With such a criterion, in situations without noise, the estimated symbols must be equal to the transmitted symbols for the desired user. At the same time, among all the wide-band global filters l_1 which cancel the interference, this criterion warrants the one with a minimal norm and leads to the structure in Figure 2 (or equation (27)). The cancelling of the interference from the symbol-band model requires then that $\hat{glo}_{1k}(f) = \delta_{k1}$, which is equivalent to $glo_{1k[n]} = \delta_{k1}\delta_n$ in discrete time representation.

The equalizer vector is then obtained by inverting the spectral matrix of aliased cross-channels:

$$\text{(28)} \qquad \underline{\hat{e}_1}^T_{zf}(f) = \underline{1}^T[\underline{\underline{\Gamma}}(f)]^{-1}$$

This is immediately derived from the equation (23) of the symbol-band model. So, the resulting MSE is formed only by the contribution of the noise, amplified by the interference cancelling process:

$$(MSE)_{zf} = (2N_0 \cdot \frac{1}{Ts}) \cdot \frac{1}{\gamma_{11[0]}} \cdot e_{11zf[0]}$$

The zero delay coefficient of the equalizer in branch "1", $e_{11zf[0]}$, is a positive real number, derived from the diagonals of the Hermitian matrix. It plays a fundamental role since it describes entirely the performance of the system. More precisely, it expresses a degradation factor for the SINR but also for the Error Probability (Pe or BER)[2] since after perfect theoretical equalization, we obtain exactly the desired symbols, only disturbed by a gaussian noise:

(29) $$(SINR)_{zf} = \frac{A^2}{(MSE)_{zf}} = (\frac{2E_b}{N_0}) \cdot \frac{1}{e_{11zf[0]}}$$

(30) $$(Pe)_{zf} = Q\left(\sqrt{(\frac{2E_b}{N_0}) \cdot \frac{1}{e_{11zf[0]}}}\right)$$

The degradation factor corresponds to the frequency-domain area of the first diagonal "block" (or frequency pattern) of the inverted aliased cross-spectral matrix:

$$e_{11zf[0]} = Ts \int_{\frac{1}{Ts}} [\underline{\underline{\Gamma}}(f)^{-1}]_{11}\, df$$

The Schwartz inequality lets us verify that $e_{11zf[0]} \geq 1$. Of course, no degradation ($e_{11zf[0]} = 1$) corresponds to an ideal situation with a single path channel and orthogonal codes.

Figure 8 gives the example of an inverted matrix $|\underline{\underline{\Gamma}}(f)^{-1}|$ in Pedestrian B. The patterns of the first line represent the frequency modulus of the equalizing filters $\hat{e}_{1k}$ for $k = 1$ to 8; the degradation factor for the desired user "1" is rather weak, computed to +1.2 dB.

Note:

From the received signal $r(t)$, there are an infinity of linear detectors l_l able to cancel interference. The linear ZF equalizer recalled here (with theoretical frequency description and therefore no constraint of finite length for the impulse responses) is optimal since it cancels interference but also minimizes, under this constraint, the noise amplification. It is not the case for a ZF detector which would consist in inverting the wide-band continuous channel, in the bandwidth of the RRC, before making the correlation with the desired code. Such a detector would generally present much worse performances.

Figure 8. Matrix $|\underline{\underline{\Gamma}}(f)^{-1}|$ example in Pedestrian B.

V.2.2. MMSE equalizer

The mean squared error is at a minimum when the quadratic form (26) of the spectral density of the error is minimized for each frequency f of the symbol band, which leads to:

(31) $$\hat{e}_{\underline{1}\,me}^{\,T}(f) = \underline{1}^T[\underline{\underline{\Gamma}}'(f)]^{-1}$$

with $\underline{\underline{\Gamma}}'(f) = \underline{\underline{\Gamma}}(f) + (\frac{N_0}{2E_b}) \cdot \underline{\underline{I}}_K$

The expression for the MMSE is then:

$$(MSE)_{me} = A^2(1 - glo_{11me[0]})$$

$$= (2N_0 \frac{1}{Ts}) \cdot \frac{1}{\gamma_{11[0]}} \cdot e_{11me[0]}$$

The inversion of the aliased cross-channels spectral matrix is realized here after adding a noise to signal $\frac{N_0}{2E_b}$ factor on the diagonal. So, as in a single user situation, the partial inversion of the MMSE limits the amplification of the noise to the frequencies where the local signal to noise ratio is bad. Moreover, since the power of the noise is not zero, the inverse of $\underline{\underline{\Gamma}}'(f)$ always exists an ill conditioned situation is impossible, contrary to the ZF case.

The global gain (at zero delay) of all the chain Tx/Rx for user "1", $glo_{11me[0]} = Ts \int_{\frac{1}{Ts}} [\underline{\underline{\Gamma}}'(f)^{-1}\ \underline{\underline{\Gamma}}(f)]_{11} df$, is a positive real less or equal to 1 (equal to 1 in a noise-free situation). The MMSE receiver has the same behavior as the ZF receiver in situations of strong $\frac{E_b}{N_0}$ and is reduced to the receiving head, i.e. to the MF, inthe extreme opposite situation. The zero delay coefficient of the equalizer is here normalized by the global gain to form the degradation factor of the SINR (≥ 1), which is expressed through:

$$(32) \qquad (SINR)_{me} = \frac{A^2 glo^2_{11me[0]}}{(MSE)_{me} glo_{11me[0]}}$$

$$= \left(\frac{2E_b}{N_0}\right) \cdot glo_{11me[0]} \cdot \frac{1}{e_{11me[0]}}$$

For strong $\frac{E_b}{N_0}$ situations, the SINR of the three receivers described in this chapter are ordered in the following manner:

$$(SINR)_{mf} \leq (SINR)_{zf} \leq (SINR)_{me} \leq \left(\frac{2E_b}{N_0}\right)$$

$$\text{when } \frac{2E_b}{N_0} >> 1$$

Then, the performances of the receiver are expected to be improved by equalization, preferentially MMSE if the noise level is not negligible.

V.2.3. Comparisons and design elements

The multi-user equalizer we have just described generalizes the classical results obtained with weakly selective channels, based on the correlation matrix of the codes at zero delay. When the temporal spread of the channel is weak compared to the symbol duration, the equalizer needs no memory dependence (white response in frequency) and is reduced to instantaneously and linearly combining the $y_{k[m]}$. In this case, the spectral matrix of aliased cross-channels, white in frequency, but non-diagonal if the coded waveforms are not orthogonal, is identical to the zero delay correlation matrix, as the frequency dimension does not appear. We then find the equalizer described by the authors of [16, 5, 6], where the coefficients combining $e_{1k[0]}$ from the branches $k = 1$ to K are given by the first line of the inverse of the zero-delay correlation matrix of the codes, equal to [1, 0, …, 0] if the codes are orthogonal.

In the presence of selective AT/RGC in the symbol band, the equalizing filters may use practically a finite temporal depth of P coefficients (with causal and anti-causal parts). We then have added the frequency dimension f, which is equivalent to independently treating several non-selective channels, with narrow bands compared to $\frac{1}{Ts}$ (second order stationarity).

This approach naturally (but approximately) provides the coefficients of the equalizer by using the Inverse Discrete Fourier Transform, and constitutes an alternative solution to temporal methods [12, 4], with little complexity.

To obtain a temporal depth of P coefficients for the equalizer, we just have to sample the symbol band with a step $\Delta f = \frac{1}{PTs}$ to form the P matrix $\underline{\underline{\Gamma}}(p \cdot \Delta f)$ and compute their inverse. The K non-causal filters $\{e_{1k[p]}\}$ used to detect the symbols of user "1" are obtained by a DFT^{-1} of the K spectral patterns of the line "1" of the inverse matrix. The filter $\{e_{11[p]}\}$ born from the diagonal is the most energetic and corresponds to the main branch. In a multi-code case, i.e. with several desired codes, we can take advantage of the hermitian symmetry of the taps: $e_{ik[n]} = e^*_{ki[-n]}$, $\forall i, k = 1 \ldots K, \forall n \in \mathbb{Z}$.

Notes:

* The impulse responses of the equalizer bank at symbol time after the MF bank are theoretically infinite. So, the choice of P should ensure a negligible temporal aliasing for the practical filters. Depending on to the weak values of the secondary far paths in the ETSI models described in Section VII, a typical temporal depth of 1 to 4 coefficients in indoor A and of 8 to 16 coefficients in Vehicular B is generally sufficient.

* Instead of preserving approximately the theoretical imposed structure described in Figure 2 (with a finite number of taps for the practical equalizer), it is also possible to compact the linear detector l_1 in a single discrete transverse fractional filter with a cadence of $Tc/2$ at input and Ts at output.

VI. Extension to multi-sensor reception

When the mobile receiver uses $L = 2$ or 3 sensors (with a spacing above $\lambda/2 = 7.5$ cm), the $i = 1, \ldots, L_t$ propagation paths are observed on the different elements $l = 1, \ldots, L$ of the antenna. The delays of one path "i" on the different sensors "l" are nearly identical according to chip time (cf note in Section II), only the complex amplitudes α_i^l are different. Thus, we define L propagation channels $h^{(l)}(\tau)$ from the model of (3), still considered here as deterministic and time invariant during the slot.

With the *l-th* sensors are also associated a received signal $r^{(l)}(t)$, a spatio-temporally white noise $n^{(l)}(t)$, a wide-band waveform for the *k-th* user $g_k^{(l)}(\tau) = (c_k * h_e * h^{(l)})(\tau)$ and transfer matrix sources/bands $\underline{\underline{G}}^{(l)}(f)$. With multi-sensor,

$\gamma_{11[0]}$ and E_b represent respectively the normalized "energy" of the waveform and the average energy per bit of the desired signal *over the L receiving sensors.*

The channel matched filter becomes spatio-temporal [1, 7, 11, 15] but similar to the multi-band dimension, the spatial dimension does not appear explicitly in the "symbol-band" model and *the digital equalizer keeps exactly the same expressions as function* of $\underline{\underline{\Gamma}}(f)$ rather than those established in single-sensor, described by equations (28, 29, 30, 31, 32). Obviously, the global diversity is concealed in $\underline{\underline{\Gamma}}(f)$ since $|\hat{h}^2|$ is replaced by $\sum_l |\hat{h}^{(l)}|^2$ in expresion (17) of the T/RGC $\hat{\gamma}_{ik}^{cont}(f)$ and thus indirectly in the formation of the a T/RGC $\hat{\gamma}_{ik}(f)$, leading to:

$$\underline{\underline{\Gamma}}(f) = \frac{1}{\gamma_{11[0]}} \cdot \sum_l \underline{\underline{G}}^{(l)H}(f)\ \underline{\underline{G}}^{(l)}(f)$$

The coherent re-combination of the spatial and frequency channels improves the orthogonality of the waveforms and therefore makes easier the system inversion operated by the joint detection equalizer. In terms of implementation, we can approximate (with finite length) the imposed structure of Figure 9, taking benefit of the discrete nature of the paths in the channel MF. Or we can implement linear receivers by means of one single fractional filtering per sensor. In the next section we are now going to measure the benefit of the multi-sensor reception.

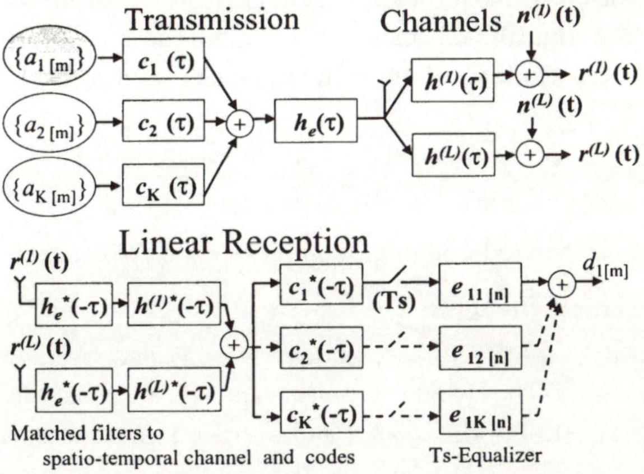

Figure 9. Multi-sensor transmission model.

VII. Average performances in indoor and vehicular environments

VII.1. Context

VII.1.1 Codes used

According to the general hypothesis, the codes $\{c_{k[q]}, q=0, \ldots, Q-1\}$ for the users k = 1, ..., K are complex with values proportional to $\{1, j, -1, -j\}$, algebraically independent and moreover, in conformity with the norm [21], algebraîcally orthogonal (cf (2)). One code is comprised of Q = *16 chips* obtained from a "channel code" $\{ch_{k[q]}, q=0, \ldots, Q-1\}$ and a "scrambling code", $\{\upsilon_{[q]}, q=0, \ldots, Q-1\}$. The channel code is particular to one user, with real binary values ($\in\{-1; +1\}$) chosen in the family of 16 orthogonal Walsh-Hadamard sequences. The scrambling code is specific to each cell, with real binary values indexed in the norm. The complex values of the final code are obtained by a rotation of $\pi/2$ from one chip to another after multiplication of the two previous codes:

$$c_{k[q]} = ch_{k[q]} \cdot \upsilon_{[q]} \cdot (j)^{q+1}, q=0, \ldots, Q-1$$

where j is the an imaginary number such that $(j)^2 = -1$.

So, with such a complex code, both real and imaginary parts of the transmitted symbols benefit by the diversity of the **I**n phase and **Q**uadrature carrier modulation. Moreover, during the transmission of one symbol, the constellation at the chip time presents no zero crossing, ensuring small envelope variations for the transmitted modulated signal, *of one given user*. The computation of the performances are related to the desired code "1" defined by:

$$ch_{1[q]} = 1\ \forall q;\ \underline{\upsilon} = [-1, +1, -1, -1, -1, +1, -1, -1, +1, -1, +1, +1, -1, +1, -1, -1]^T$$

This code "1" is then reduced to the complex version of the scrambling code.

VII.1.2. Propagation channels

Models described by ETSI [20] in "Indoor A" (IA) and "Vehicular B" (VB) environments are used. The channel impulse responses are constituted by 6 paths with relative delays (to the first paths) and average power, summarized in the following table:

IA (nsec)	0	50	110	170	290	310
(dB)	0	– 3	– 10	– 18	– 26	– 32
VB (μsec)	0	0.3	8.9	12.9	17.1	20.0
(dB)	– 2.5	0	– 13	– 10	– 25	– 16

The temporal spreading is very weak in IA, slightly superior to one chip time, *Tc*; on the contrary, the temporal spread is important in VB, around 5 *Ts*, which will permit equalizer performances in two very different situations. The propagation delays are quasi-fixed (variation of $\pm 3\%$). The complex amplitudes α_i^l of the paths are random, distributed as with a Rayleigh law for the modulus and uniform law between 0 and 2π for the phases. We suppose the complex amplitudes uncorrelated between two paths (index "*i*") and from one sensor to another (index "*l*"). This last hypothesis, more often admitted [10], is certainly a bit optimistic for the inter-sensor spacing looking at the mobile; we will come back to this point in the performance analysis.

VII.1.3. Average performances (random channels)

We are going to evaluate, for one given model of channel, the performances of the equalizers from theoretical formulas (24, 29, 30, 32) developed in Section V. The frequency integrals are approximated by the rectangle method, with a frequency step $\Delta f = \frac{1}{32Ts}$ and atemporal oversampling of *Tc*/8 to form the delays of the paths. These performances take the form $(SINR) = (\frac{2E_b}{N_0}) \cdot \frac{1}{\beta_I}$ where β_I *represents the degradation* on the SINR *caused by interference* for the Matched Filter, Zero-Forcing and MMSE receiver. For the Zero-Forcing equalizer, β_I also represents the degradation of the BER.

We recall that these performance formulae suppose perfect knowledge of the channel and perfect timing synchronization, they give then upper bounds which will inevitably be degraded in real situations. Moreover, they are only valid for a deterministic and time invariant impulse response, which does not correspond to IA and VB models since they are random, mainly from the complex amplitudes α_i^l. Rather than calculating the performances for one specific observed sample of the path amplitudes (supposed then deterministic, equal for example to their quadratic average values), we are going to give *average performances* for the whole possible occurence of impulse response. This classical process [17, 10] may be used to represent a practical situation with slowly time variant channel: one occurence of α_i^l is associated with the channel impulse response at a given instant. The channel variation is slow enough to be considered fixed during the occurence and perfectly estimated. *So, the interference term* β_I, *as with the energetic gain of the global waveform* $\gamma_{11[0]}$ *become random variables*. We are now interested in the expected values of the SINR inverse (named INSR) and of the Error Probability (for ZF), computed from a very large sampling number:

$$\overline{(INSR)} \triangleq E\left(\frac{1}{SINR}\right) = \left(\frac{N_0}{2\overline{E}_b}\right) \overline{(\beta_I \cdot \beta_f)} \tag{33}$$

(34) $$(\overline{Pe})_{zf} = E\left\{Q\left(\sqrt{(\frac{2\overline{E}_b}{N_0})\cdot\frac{1}{\beta_I}\cdot\frac{1}{\beta_f}}\right)\right\}$$

where $\beta_f \triangleq \frac{\overline{\gamma}_{11[0]}}{\gamma_{11[0]}} = \frac{\overline{E_b}}{E_b}$ *expresses the amplitude fading*, the on-lining representing the expected or average value. The degradation is due to two different phenomena: the interference, through β_I, and the amplitude fading (spreading of the range of the probability density function of the normalized energy $\frac{\overline{E_b}}{E_b}$, through β_f).

In a order to obtain the average degradation caused only by the interference, a *situation without fading* is artificially introduced. In this situation, from one occurrence to another, the waveform is still random, but its energetic gain $\gamma_{11[0]}$ is maintained at a constant value $\hat{\gamma}_{11[0]}$, which fixes the receive average bit energy to $E_b = \overline{E}_b$ and makes deterministic $\beta_f = \overline{\beta}_f = 1$.

In a situation with fading, only the BER performances in Zero-Forcing are plotted since the $\overline{INSR}$ curves are practically derived from those of the situation without fading, by an additional weighting factor $\overline{\beta}_f$, superior or equal to 1. This last observation is explained by equation (33) and by the weak correlation between the fading term β_f and the interference term β_I, as shown in Figure 10 which plots the correlation coefficient related to the number of users, from the model VB. $\overline{\beta}_f$ *is thus an indicator of the degradation due to the fading,* even if it is well known that the average energetic indexes are not sufficient to express quantitatively the Error Probability. Without interference ($\beta_I = 1$), with a single-path channel subject to Rayleigh fading, and with uncorrelated envelopes from one sensor to another, Pe asymptotically varies [17] with $(\frac{\overline{E_b}}{N_0})^{-L}$. The slope of the BER curve (with classical Logscales) is imposed by the diversity order, equal to the number of sensors in this basic example.

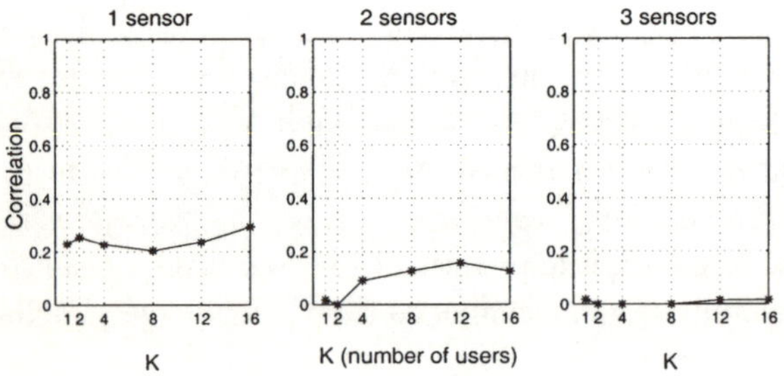

Figure 10*. Correlation coefficient between β_I and β_f, computed with the VB channel model.*

With any random channel, we may still define an equivalent diversity order N_{eq} relative to a Rayleigh fading, as proposed in [9], from the variation coefficient cv of the power received:

$$N_{eq} = \frac{1}{cv^2} \text{ with } cv = \sqrt{\frac{\overline{E_b^2} - \overline{E_b}^2}{\overline{E_b}}}$$

We then dispose of two indexes to measure the depth of the fading: cv and $\overline{\beta_f}$. With respect to a Rayleigh fading with N independent diversity branches (with same power), it can be shown that:

$$\overline{\beta_f} = \frac{N^N}{N-1}\left(\frac{1}{N}\right)^{(N-1)}$$

In order to facilitate the interpretation of the results, we indicate in the following table the values (in decibel) of these two indexes for a number of diversity orders:

N_{eq}	1	2	3	5	8	$+\infty$
cv (dB)	0	–1.5	–2.3	–3.5	–4.5	$-\infty$
$\overline{\beta_f}$(dB)	$+\infty$	3	1.76	0.96	0.57	0

VII.2. Performance analysis

VII.2.1. Vehicular without fading

Figure 11 presents the results in VB with 1, 2, 3 sensors for a situation without fading. Each elementary cell of the figure shows the results superimposed for 1, 2, 4, 8, 12 and 16 users in continuous line, from bottom to top. The dotted lines give the theoretical ideal performances obtained with a single path channel without fading.

Concerning the $\overline{(INSR)}$, we first observe a great improvement (diminution of $\overline{\beta_I}$) in MMSE or ZF (equivalent to MMSE for strong $\frac{E_b}{N_0}$) as compared with the single matched filter. For the MF, the $\overline{(INSR)}$ curves reach a floor, for strong $\frac{E_b}{N_0}$ and for the large values of K, because of the interference. The improvement due to the equalization is emphasized with 2 or 3 sensors, which brings it almost back to ideal performances.

Concerning the BER in ZF, there is an important loss, around 15 dB at Pe = 10^{-3} when the saturation of the number of users (K = 16) is approached. The multisensor reception gives a new degree of freedom and reduces the degradation to just 1 dB with 3 sensors.

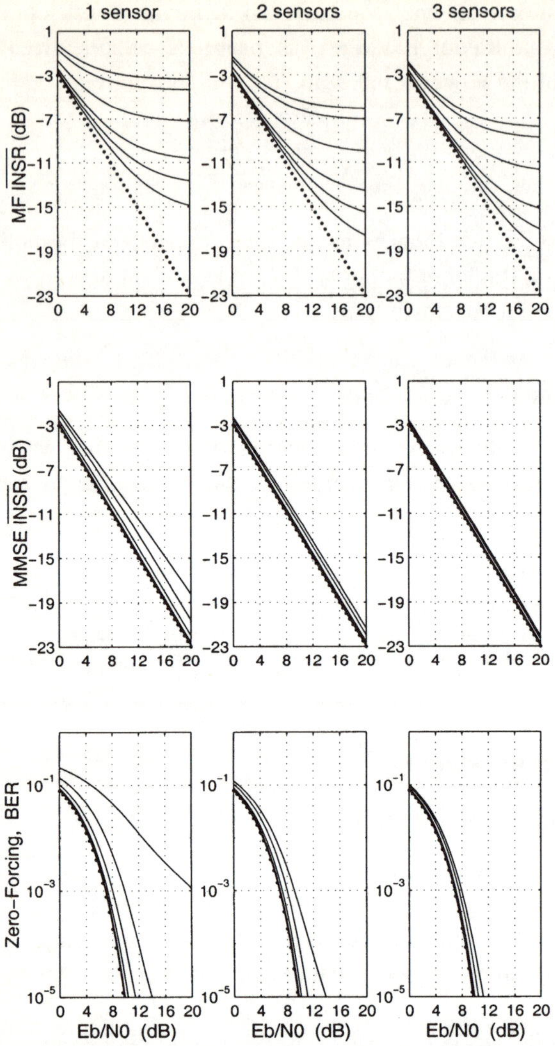

***Figure 11**. VB performances without fading.*

VII.2.2. Vehicular with fading

Figure 12 presents the results for a ZF receiver in a fading situation (top of the figure). The upper dotted lines correspond to the performance with a single-path channel subject to Rayleigh fading with only one receiving sensor, i.e. without diversity. The importance of the fading process is embodied by the bottom plots which indicate the scattering of the probability density of the channel energy, related to its average value. We also have mentioned the indexes cv and $\overline{\beta}_f$

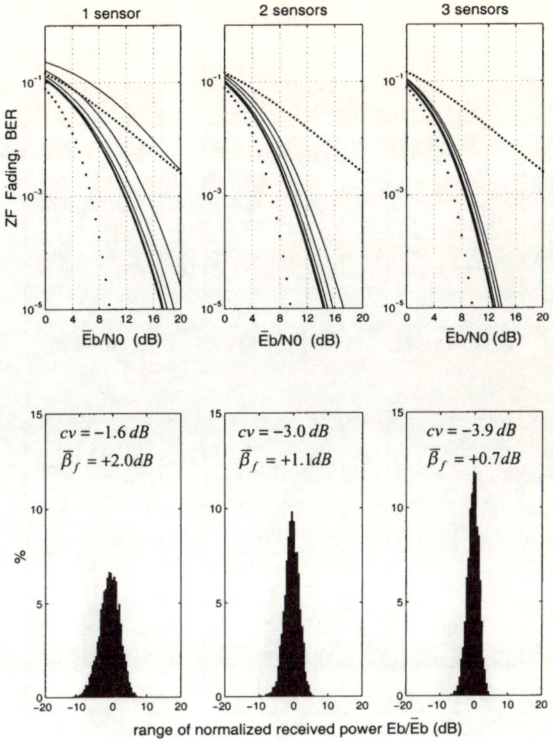

***Figure 12**. VB performances with fading.*

We observe with a single-sensor a degradation of 3 to 5 dB compared to the curves of BER without fading the situation. The very important degradation above 18 dB for 16 users is reduced to 3 dB with 3 sensors.

We notice, for one user and no spatial diversity, that the BER sensitivity is better than in a single path fading situation. This is naturally explained by the path diversity (resolved paths) of the Vehicular environment. Indeed we have for a single-sensor $cv = -1.6$ dB and, $\bar{\beta}_f = 2$ dB, which already corresponds to a diversity order N_{eq} of between 2 and 3. This order approaches 5 and 7 with 2 and 3 sensors.

VII.2.3. Indoor with fading

It is interesting to compare the fading situation previously noticed with those observed in the IA environment (Figure 13) where there is almost no path diversity and very few interferences in the MF outputs. With a single sensor, the error probability is nearly equivalent to the one path fading situation for any number of users, expressing weak β_I and strong $\bar{\beta}_f$. The spatial diversity is then

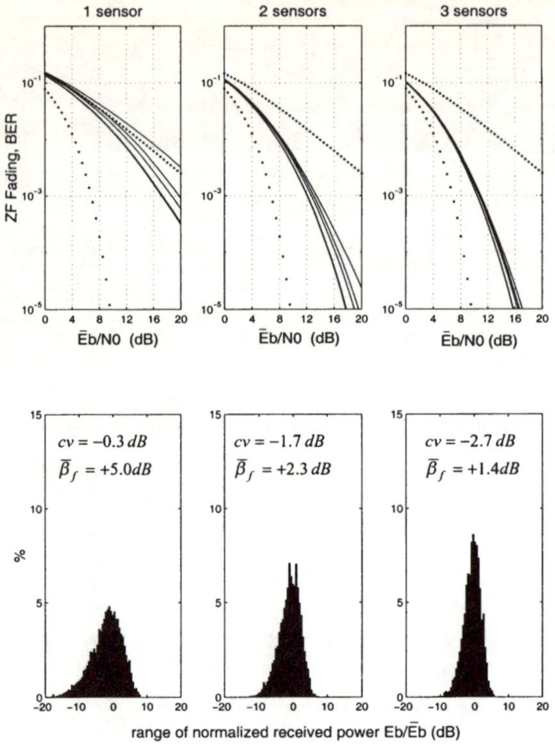

Figure 13. *IA performances with fading.*

required, even if K is small, in order to decrease the range of the amplitude fading: for one single sensor, $cv = -0.3$ dB and $\overline{\beta}_f = 5$ dB, which corresponds to a diversity order N_{eq} of very little, slightly above 1. This order approaches 2.5 and 4 respectively with 2 and 3 sensors.

This improvement would be less significant if the paths were correlated from one sensor to another. Nevertheless, the Indoor environment, which mostly needs spatial diversity, presents a very favourable situation in downlink reception on the mobile [18, 9], due to the proximity between the scattering objects and the receiver. In this case, one path can be modelled by a lot of micro-paths arriving at the mobile sensor with quasily the same delay. Depending on to the model developed in [18] for one macro-path, the correlation between any field component envelope measured at two separate points on the mobile is shown to be weak with a spatial separation above $\frac{\lambda}{2}$. To the contrary, the reception situation on the base stationin "uplink" would be very different because of a geometrical asymmetry and would require a spacing above 50 λ for obtaining weak correlations.

In conclusion, a coherent multi-element antenna in reception improves the transmission by fighting against both frequency selectivity and amplitude fading.

Note: In terms of link budget, the presented performances should benefit from an additional improvement of $10log(L)$ dB due to the antenna gain. All the curves were plotted in the function of the global energy of all the sensors.

VIII. Conclusion

This study has permitted, by taking a frequency interpretations approach, to recall non-exhaustively the linear detection structures in the downlink of a TD-CDMA system. We have formulated the expressions and performances of this detector from the representation in the symbol-band, based on the spectral matrix of the aliased cross-channels. Finally, we have applied the performances expressions to the VB and IA environment models. These performances constitute upper bounds for real situations because of the inaccuracy in channel estimation and in synchronization operation.

From the VB model, frequency selective, we have first illustrated the benefit of the joint detection as compared with the simple desired waveform matched filter, especially when the code margin is little. We have also underlined the clear performance improvement for a coherent reception with 2 or 3 sensors. In this selective channel, the multi-sensor reception permits the decrease of interference and to facilitation of the inversion performed by the equalizer.

From the IA model, weakly selective but also very poor in temporal diversity (non-resolved paths), we notice a spectacular improvement of performances due to the diversity gain of the multi-sensor, reducing the range of amplitude fading.

So, multi-sensor reception on the mobile is always recommended to fight efficiently the two different types of distortion introduced by the channel: frequency selectivity and fading process.

REFERENCES

[1] Van Etten (W.), Maximum likelihood receiver for multiple channel transmission systems. *IEEE Transactions on Communications,* pp. 276-283, (Feb. 1976).

[2] Turin (G.L.), Introduction to spread-spectrum antimultipath techniques and their application to urban digital radio, *Proceedings of the IEEE*, vol. 68, n° 3, pp. 328-353, (March 1980).

[3] Pickholtz (R.L.), Schilling (D.L.), Milstein (L.B.), Theory of spread-spectrum communications – A tutorial. *IEEE Transactions on Communications,* vol.com-30, n° 5, pp. 855-884, (May 1982).

[4] Jung (P.), Blanz (J.), Joint detection with coherent receiver antenna diversity in CDMA mobile radio systems. *IEEE Transactions on Vehicular Technology,* vol. 44, n° 1, pp. 76-88, (Feb. 1995).

[5] Madhow (U.), Honig (M.L.), MMSE interference suppression for direct sequence spread-spectrum CDMA. *IEEE Transactions on Communications,* vol. 42, n° 12, pp. 3178-3188, (Dec. 1994).

[6] Wang (X.), Poor (H.V.), Blind multiuser detection: a subspace approach. *IEEE Transactions on Information Theory,* vol. 44, n° 2, pp. 677-690, (March 1998).

[7] Balaban (P.), Salz (J.), Optimum diversity combining and equalization in digital data transmission with applications to cellular mobile radio-Part I: Theorical considerations. *IEEE Transactions on Communications,* vol. 40, n° 5, pp. 885-894, (May 1992).

[8] Bourennane (S.), Faure (B.), Lacoume (J.L.), Traitement d'antenne pour des sources bande large. *Annales des Télécommunications,* vol. 45, n° 7-8, pp. 384-392, (July 1990).

[9] Diouris (J.F.), Zeidler (J.), Buljore (S.), Space-path diversity in CDMA using a compact array. *Annals of Telecommunications,* vol. 53, n° 11-12, pp. 425-434, (1998).

[10] Buljore (S.), Diouris (J.F.), Saillard (J.), Étude théorique d'un égaliseur multi-capteur associé au critère EQM pour le canal radiomobile (modèle GSM), GRETSI, *Juan-les-Pins,* pp. 557-560, (Sept. 1995).

[11] Vila (P.), Pipon (F.), Pirez (D.), Fety (L.), Filtrage adapté multidimensioonel pour l'égalisation d'un canal sélectif en fréquence et brouillé. *GRETSI, Juan-les-Pins,* pp. 553-556, (Sept. 1995).

[12] Piceonnat (Y.), Alternative solutions for joint detection in TD/CDMA multiple access scheme for UMTS. *SPAWC Anapolis,* pp. 329-332, (May 1999).

[13] Ros (L.), Jourdain (G.), Arndt (M.), Multi-sensor reception for TDD-UMTS mobile terminal in indoor and pedestrian environments. *11th IEEE International Symposium on PIMRC, London,* pp. 1221-1225, (Sept. 2000).

[14] Ros (L.), Réception multi-capteur pour un terminal mobile TDD-UMTS. *Report LIS n°8/2000,* (May 2000).

[15] CAPELLANO (V.), Apport des traitements spatio-temporels pour les transmissions numériques en présence de trajets multiples. Application aux communications acoustiques sous-marines, *Phd thesis, INPGrenoble,* (May 1998).

[16] VERDU (S.), Multiuser detection. *Cambridge University Press,* (1998).

[17] PROAKIS (J.G.), Digital communications. *McGraw-Hill, third edition,* (1995).

[18] JAKES (W.C.), Microwave mobile communications. *IEEE PRESS, New-York,* (1974).

[19] MARCOS (S.) et al: Les méthodes à haute résolution, traitement d'antenne et analyse spectrale. *Collection Traitement du Signal, Hermes,* (1998).

[20] ETSI: chapter B.1.4.2, Channel impulse response model, UMTS 30.03, TR101-112 V3.2.0, (April 1998).

[21] 3GPP: Technical specification: spreading and modulation (TDD). *TS25.223, V3.2.0,* (March 2000).

Chapter 5

Smart-antenna space-time UMTS uplink processing for system capacity enhancement

T. Neubauer

Technical University, Vienna, Austria

E. Bonek

Technical University and FTW, Vienna, Austria

I. Introduction

The European third generation mobile communication system is known as the Universal Mobile Telecommunications System (UMTS). It will enable tomorrow's wireless Information Society, to deliver high-value broadband information, commercial and entertainment services to mobile users. UMTS will play a key role in creating the future mass market for high-quality multi-media communications that will approach 2 billion users worldwide by the year 2010 [1].

The main advantage of third generation mobile systems will be mixed services with variable data rates. The customer profile of a 3G system will consist of a large number of low-data-rate and voice users with the concurrent service to a high-data-rate user population.

The frequency division duplex (FDD) mode of UMTS is based on a Wideband Code Division Multiple Access (W-CDMA) technique. CDMA systems are interference limited in nature and suffer severely from the near-far effects because all users operate in the same frequency band simultaneously. This is essentially more critical in the uplink of UMTS where all users are potential interferers. The problems due to the near-far effect are reduced in the downlink because of the better orthogonality of the individual signals [2].

Since higher bit-rate services experience lower processing gains, they have to be sent with higher power levels in order to fulfil the quality requirements at the receiver. This causes greater interference at the base station (BS) antenna and an intensification of the near-far problem. Hence, the capacity will be reduced due to the overall interference increase.

I.1.1 Rake receiver structures

Radio propagation in the land mobile channel is characterized by multiple reflections, diffraction and attenuation of the signal energy. These are caused by natural obstacles such as buildings, vehicles, hills, and the like, resulting in multi-path propagation. The signal energy may arrive at the receiver over many clearly distinguishable time instants. The delay profiles extend typically up to 1 or 2 μsec in urban and suburban areas [4]. In the FDD mode of UMTS, the chip duration is 0.26 μsec. If the time difference of the multi-path components is at least 0.26 μsec, the W-CDMA receiver can separate these multi-path components and combine them coherently (*rake* them up) to obtain multi-path diversity.

Rake receiver structures are well known from second generation CDMA systems like IS 95 where they are implemented. Available literature on rake receiver structures is vast and covers capacity evaluation as well as implementation aspects [8, 12, 26, 28, 29, 30].

I.1.2 Smart-antennas

Smart- or adaptive-antennas have been recently applied to improve the capacity and the performance of second generation wireless mobile communication systems [17, 20, 21, 33]. In general, smart-antennas reduce the effects of multipath fading and improve the signal-to-noise-plus-interference ratio (SNIR). Additional advantages such as channel capacity increase in urban areas, coverage extension in rural areas and spectrum efficiency increase in general can be obtained with the aid of smart-antennas at the base station. Drawbacks are cost, and both hardware and software complexity.

In this paper we investigate the effects/benefits of smart-antennas in the FDD uplink by means of UMTS system-level simulations. A simplified model for smart-antennas in a multi-service UMTS scenario is shown in Figure 1. We assume that the cell serves several low data rate users and a few high data rate users. The latter are indicated by mobile terminals with large screen and keyboard.

The high data rate users heavily interfere with the far-off user equipment (UE).[1] With smart-antennas the interference at the BS can be reduced for this UE by placing broad nulls in the direction of the high data rate users. Due to this, both the coverage range (limited by the TX power of the desired UE) and the received signal quality will increase.

Alternatively, smart-antennas will increase the received SNIR. Power control algorithms designed to keep the reception quality at the same level will then reduce the transmit power of the UEs. This will lead to a overall decrease in the multiple-access interference (MAI), which, in turn, leads to higher system capacity.

1. Note that there are some similar notations in the standard. MS (mobile station), ME (mobile equipment) and UE (user equipment). They sound similar, but they have different meaning [35].

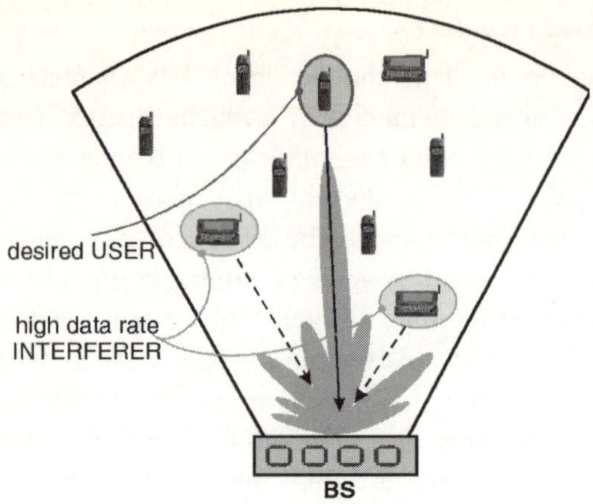

Figure 1. Smart-antenna patterns in a multi-service UMTS system with high-data-rate interferers and desired low-data-rate users.

Taking all this into account, smart-antennas promise a clear advantage in multi-service traffic scenarios in comparison with non-adaptive antenna techniques. It is generally accepted that the adaptive antenna technique will be a key enabler for the success of UMTS [34].

II. Data model

In the following we present the channel model used, signal model and the overall system model.

II.1. Channel model

An important part of the evaluation process of a system employing smart-antenna arrays is the channel modeling. For system capacity investigations, flat fading channel models are widely used [22, 23]. This is a gross simplification of the nature of mobile radio propagation channels and does not capture the detailed information of state of the art channel modelling [4]. Hence, we study the impact of a channel model that includes both the spatial and the temporal distribution of the energy reaching the receiver. Moreover we compare this with the results obtained under flat fading assumptions.

II.1.1. Flat fading channel

The flat fading assumption implies that the delay spread introduced by the channel is much smaller than the time duration of the transmitted symbols. In our idealized flat fading channel the delay spread is zero and hence all incoming waves are received with delay $\tau = 0$.

It is well accepted that the received signal at the BS originates from scatterers within a disk of radius r centred R metres from the BS around the UE [24]. Measurements in various urban environments have shown that the spatial behaviour of the incoming signals at the BS can be modelled by a main direction of arrival (DOA) in the azimuth and a corresponding angular spread [5].

The path loss is modelled based on the proposed models for UMTS system simulations presented in [2] for pre-defined environments. Path loss correlation between two up-link connections from a single UE to different BSs is expressed by a correlation coefficient of 0.5 [10]. The log-normal large-scale fading is modeled with a mean of $\mu_{ln} = 0$ and a standard deviation of $\sigma_{ln} = 6$ dB.[2]

II.1.2. Spatio-temporal channel model

Within the European research initiative COST 259 [4] several channel models have been developed. They are aimed at UMTS and HIPERLAN, with particular emphasis on adaptive antennas and directional channels. They have been introduced in the third generation standardization process by 3GPP [3], and they are used here in the following way.[3]

The path loss again is modelled based on the proposed models for UMTS system simulations presented in [2] for pre-defined environments. Path loss correlation and the log-normal large-scale fading are modelled in the same way as in the flat fading case.

In order to test advanced antenna array systems, the main focus of the channel model has to be on the azimuthal and temporal properties. The statistics of the delay and azimuth of the impinging waves as well as their expected power are especially of great interest. According to [4] and [5] we define a Laplacian shape for the Azimuth Power Spectrum (*APS*) seen at the BS antenna. For the Delay Power Spectrum (*DPS*) we assume the widely used one-sided exponential decaying function [5]. Both can be written as

(1) $$APS(\phi) \propto \exp\left(-\sqrt{2}\,\frac{|\phi|}{\sigma_\phi}\right)$$

(2) $$DPS(\tau) \propto \exp(-\tau / \sigma_\tau), \text{ for } \tau > 0$$

where ϕ, σ_ϕ, t and σ_τ denote the azimuth towards the UE seen at the BS, the azimuth spread, the delay and the delay spread, respectively. For typical urban environments the azimuth power delay spectrum can be written as

(3) $$P(\phi, \tau) \propto APS\,(\phi)\,DPS(\tau)$$

2. Enlarging the standard deviation to $\sigma_{ln} = +10$ dB has negligible impact on the results.

3. Within 3GPP [3] only thee temporal domain is considered till now.

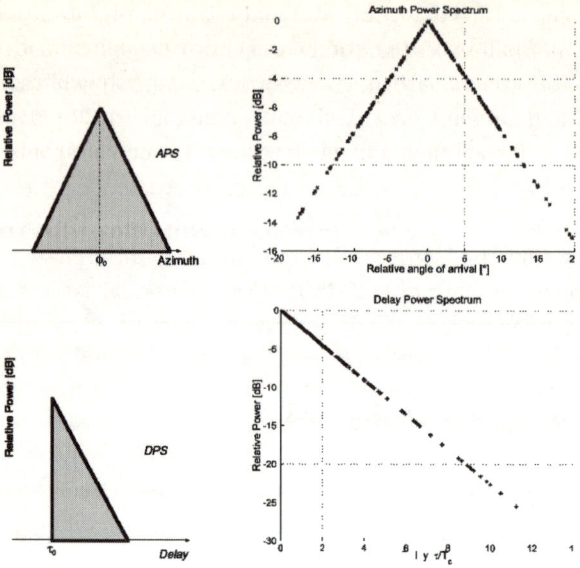

Figure 2. Assumed APS and DPS (according to the 3GPP-COST259 proposal). $\sigma_\tau = 0.5\mu sec$, $\tau_0 = 0\ \mu sec$ *and* $\sigma\phi = 8°$ *are assumed.*

In our model we assume a tapped delay line model for a typical urban scenario. We consider a single scattering cluster and L impinging multipath-components. The delay spread is $\sigma_\tau = 0.5$ μsec and the start of the cluster is at $\tau_0 = 0$ μsec [3]. For the azimuth domain we assume a spread of $\sigma_\phi = 8°$ around ϕ_0, where ϕ_0 corresponds to the azimuth toward the UE. In Figure 2 both APS and DPS are shown schematically and for the parameters above.

II.2. System model

Following the results of [9], 19 sites for single cell investigations have to be considered in the system model. For each site we assume 3 sectored cells. Hence, a total of 57 cells altogether has been taken into account.

An example for a single element pattern, which approximates a commercial UMTS antenna pattern, is shown in Figure 3. We assume that it is identical for all antenna elements.

The main system and simulation parameters are summarized in Table I.

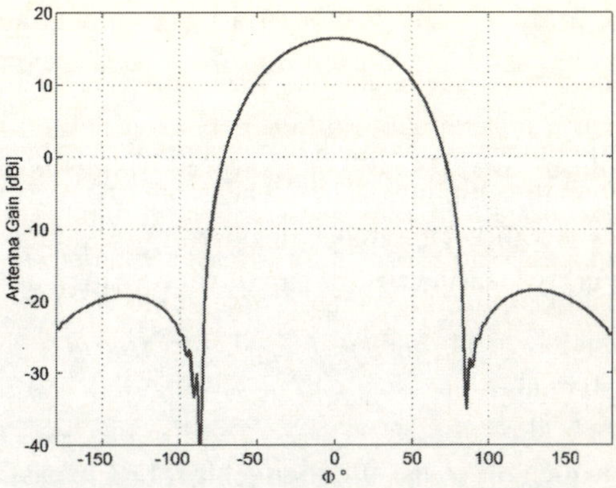

***Figure 3**. Single-element radiation pattern.*

***Table I**. Main simulation parameters.*

Number of cells	57
Inter-BS distance	1000 m
Background noise floor	– 105 dBm [36]
Path loss model	Macro cell [2]
Log-normal large-scale fading	$\mu_{ln} = 0, \sigma_{ln} = 6$ dB
Path loss correlation	0.5
Max. UE TX power	21 dBm
TPC dynamic range	80 dB
Power control error	$\mu = 0, \sigma_p = 0.5$ dB
No. of antenna elements	1 and 4
Inter-element spacing	$d = \lambda/2$
Number of impinging rays	10
Delay spread	$\sigma_\tau = 0.5\mu$sec
Azimuth spread	$\sigma_\phi = 8°$

II.3. Signal Model

Following the notation of Bello [6], the *azimuth-delay spread function* of the channel for each path l from UE k to BS ζ is given by

$$(4) \quad h_{l,k,\zeta}(\phi,\tau) = \sqrt{P_{l,k,\zeta}} \cdot e^{j\theta_{l,k,\zeta}} \cdot \delta(\phi - \phi_{l,k,\zeta}, \tau - \tau_{l,k,\zeta})$$

where $\sqrt{P_{l,k,\zeta}}$ and $\theta_{l,k,\zeta}$ denote the amplitude and the phase of the *l-th* received signal path at a single antenna element impinging with delay $\tau_{l,k,\zeta}$ and azimuth $\phi_{l,k,\zeta}$. $\delta(\phi, \tau)$ is the Dirac delta function. The array response vector for a single UE k at BS ζ for the delay τ is thus given as

$$(5) \quad \mathbf{a}_{l,k,\zeta}(\tau) = \sum_{l=1}^{L} \left\{ \int_0^{2\pi} \mathbf{c}_{l,k,\zeta}(\phi)\, h_{l,k,\zeta}(\phi,\tau)\, d\phi \right\}$$

where $c_{l,k,\zeta}$ is the array steering vector of a wave impinging from an azimuth direction $\phi_{l,k,\zeta}$ at an M-element uniform linear array with an inter-element spacing d.

In the FDD mode of UMTS, the time a wavefront takes to pass through the array is much smaller than the chip interval T_c. Therefore, the narrowband assumption for antenna arrays is valid. This makes it possible to model the time delays as phase shifts. The array steering vector can thus be written as

$$\text{(6)} \qquad \mathbf{c}_{l,k,\zeta}(\phi) = [1, e^{-j\frac{2\pi d}{\lambda}\sin\phi_{l,k,\zeta}}, ..., e^{-j\frac{2\pi d}{\lambda}(M-1)\sin\phi_{l,k,\zeta}}]^T$$

where λ is the wavelength of the impinging signal. If we assume that we know perfectly the channel, the spatial covariance matrix of the k-th user in the cell ζ at a certain delay τ becomes:

$$\text{(7)} \qquad \mathbf{R}^{(s)}_{k,\zeta}(\tau) = \mathbf{a}_{k,\zeta}(\tau)\,\mathbf{a}^H{}_{k,\zeta}(\tau)$$

In the same way we define the spatial interference and noise covariance matrix $\mathbf{Q}^{(s)}_{k,\zeta}(\tau)$ as

$$\text{(8)} \qquad \mathbf{Q}^{(s)}{}_{k,\zeta}(\tau) = \sum_{l \neq k}^{k} \mathbf{R}^{(S)}{}_{l,\zeta}(\tau) + \sigma^2_N \mathbf{I}$$

where σ^2_N and $\boldsymbol{I}$ denote the noise variance and $M \times M$ the identity matrix, respectively.

In the case of the flat fading channel model, we include an angular spread. Borrowing the technique from [25], we derive the spatial covariance matrix of the k-th user at a certain instant in time as

$$\text{(9)} \qquad \tilde{\mathbf{R}}_{k,\zeta}(t) = \mathbf{R}_{k,\zeta}(t) \odot S(\phi_{k,\zeta}, \sigma_{k,\zeta})$$

with

$$\text{(10)} \qquad [\mathbf{S}(\phi_{k,\zeta}, \sigma_{k,\zeta})]_{pq} = e^{-2[\pi d(p-q)]^2\sigma^2_{k,\zeta}\cos^2(\phi_{k,\zeta})}$$

where $\odot$ and $[.]_{pq}$ denote the Schur-Hadamard element-by-element matrix product and the pq-th element of a matrix, respectively. $\sigma_{k,\zeta}$ indicates the angular spread of the corresponding DOA $\phi_{k,\zeta}$. Applying (9) to the interference covariance matrix we get

$$\text{(11)} \qquad \tilde{\mathbf{Q}}_{k,\zeta}(t) = \mathbf{Q}_{k,\zeta}(t) \odot S(\phi_{k,\zeta}, \sigma_{k,\zeta}).$$

III. Space-time processing

Space-time processing in third generation systems can be seen as an evolution of more "traditional" receiver structures. Space-time processing combines time-only and space-only processing. Time-only processing corresponds to equalizers that use a weighted sum of signal samples (RAKE receiver), and space-only processing corresponds to simple beamforming that uses a weighted sum of antenna outputs. Examples for available literature on space-time processing are [12, 13, 19, 27, 28, 31].

A block diagram of our space-time BS receiver structure is shown in Figure 4. In the literature, this receiver structure is also named the "decoupled space-time rake", the "beamformer rake", the "2D-rake" and the "vector Rake – single beamformer".

III.1. Spatial processing

For the implementation of combining algorithms of the multiple antenna signals, it is important to know whether the spatial interference and noise covariance matrix is diagonal or not [14, 11], meaning the interference is *white* or not. In the case of a large number of users in the system, spatial whiteness can be generally expected [7, 8]. The UMTS radio network controller (RNC) will reduce the transmit power level of each individual UE to a minimum, even in the case of a few high-data-rate users. Hence, singular interferers with high power levels, which would be represented by high off-diagonal correlation, do not occur.

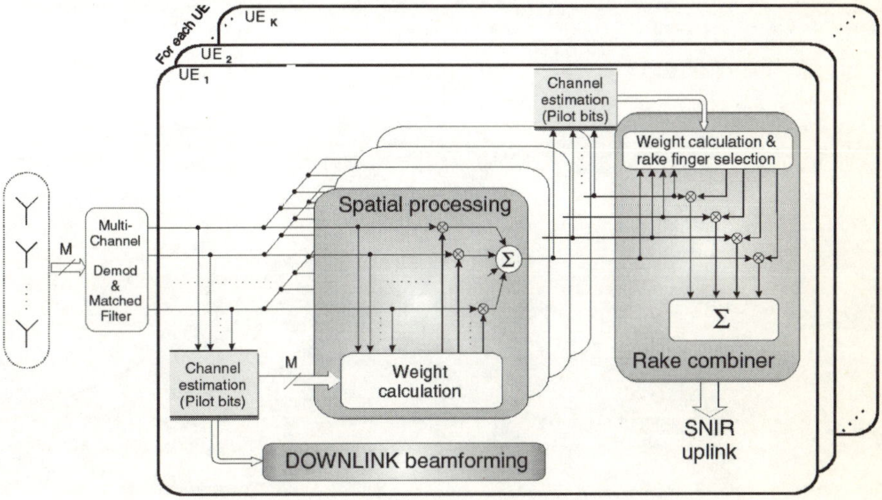

***Figure 4**. BS space-time receiver structure.*

We investigated whether interferer whiteness is fulfilled or not. An example of a long-term interference and noise covariance matrix with only few high-data-rate users is given in Figure 5. We observe that the covariance matrix has a diagonal shape indeed. But we also see that the spatial covariance matrix is not stringently *diagonally dominant* [32] and hence spatial whiteness is not given per se in a multi-service environment. This corroborates the intuition that single "very" high data users contradict the assumption of a spatially white covariance matrix.

The general problem in spatial processing is to find a weight vector for each user k in order to maximize the signal-to-noise-plus-interference ratio SNIR γ_k in the uplink.

$$(12) \qquad \gamma_k = \left(S/N+I\right)_k = \max_{\mathbf{w}_1..\mathbf{w}_k} \frac{\mathbf{w}_k^H \mathbf{R}_k^{(s)} \mathbf{w}_k}{\mathbf{w}_k^H \mathbf{Q}_k^{(s)} \mathbf{w}_k}$$

Without making any assumptions about the spatial covariance matrix behaviour, the solution for this problem is to take the generalized eigenvector corresponding to the largest eigenvalue $[\boldsymbol{R}^{(S)}{}_{k,\,\zeta},\ \boldsymbol{Q}^{(S)}{}_{k,\,\zeta}]$ [15]. This well-known beamforming concept corresponds to maximizing the signal-to-noise-plus-interference ratio SNIR in the uplink [14] and the estimated ratio of signal to interference power for the *k-th* UE. From

$$(13) \qquad \mathbf{w}_{k,\,\zeta}^{opt,\,S}(\tau) = \underset{\mathbf{w}_{k,\,\zeta}}{\operatorname{argmax}} \frac{\mathbf{w}_{k,\,\zeta}^H \mathbf{R}^{(S)}{}_{k,\,\zeta}(\tau)\,\mathbf{w}_{k,\,\zeta}}{\mathbf{w}_{k,\,\zeta}^H \mathbf{Q}^{(S)}{}_{k,\,\zeta)}(\tau)\,\mathbf{w}_{k,\,\zeta}}$$

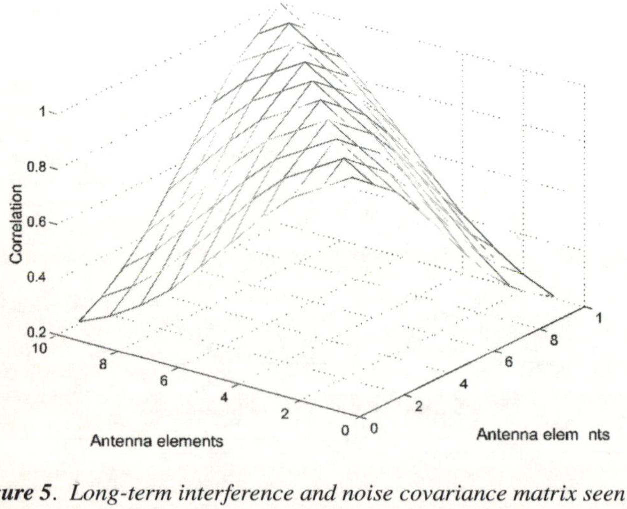

Figure 5. *Long-term interference and noise covariance matrix seen at the antenna elements of the central* BS.

we obtain the optimum spatial weights for the *k-th* UE in cell ζ as a function of the delay τ. Since **R** contains all signal information, the received signal power after spatial processing for the *n-th* delay tap is given by

$$z_{k,\zeta}(\tau_n) = \mathbf{w}_{k,\zeta}^{opt,S}(\tau_n))^H \mathbf{R}_{k,\zeta}^{(S)}(\tau_n)\,\mathbf{w}_{k,\zeta}^{opt,S}(\tau_n) \tag{14}$$

III.2. Temporal processing

Applying optimal spatial weights to the receiver structure shown in Figure 4, we obtain optimum SNIRs for each delay tap at the rake receiver input for each individual user.

For the temporal consideration we assume a wide-sense-stationary (WSS) channel with uncorrelated scattering (US) only. The complete short-term channel estimate in time can be obtained by correlation with the pilot sequence, which is part of the control channel in the UMTS FDD mode. The consequence of the US condition is that the contributions of the temporal taps selected for further processing steps are uncorrelated.[4]

An example for a temporal signal covariance matrix is given in Figure 6.

Since we only have a limited number N, of rake fingers, the taps with the highest signal power are selected. As mentioned above, we can assume temporal whiteness of the interfering signals. Thus, the optimum output of the rake receiver can be achieved by maximum ratio combining (MRC) of the selected fingers [14]. The weights for the rake fingers are thus given by

$$\mathbf{w}_n = (\mathbf{h})_n^{(T)})^* \tag{15}$$

where $\boldsymbol{h}_n^{(T)}$ and $(\cdot)^*$ denote the temporal channel impulse response and complex conjugate, respectively.

IV. Simulations

For investigations of system behaviour, we now have to consider a number of cells. As already mentioned, we consider 19 3-sectored sites, leading to a total of 57 cells. In doing so, we correctly estimate the inter-cell interference at the inner site and thus we are able to come up with reliable conclusions at the end [9].

Taking the whole network into account, we calculate the interference seen at the BS antenna for each user. Since we now have a total number of $\hat{K}$ users in the entire network, we easily find from (8)

4. Two adjacent taps are usually not completely uncorrelated due to filtering and oversampling. Analysis in [16, chapter 29.3 "The selective rake receiver"] has shown that the correlation is in the range of 0.02 for raised cosine pulse shapes. Thus, we *assume* two adjacent taps to be uncorrelated.

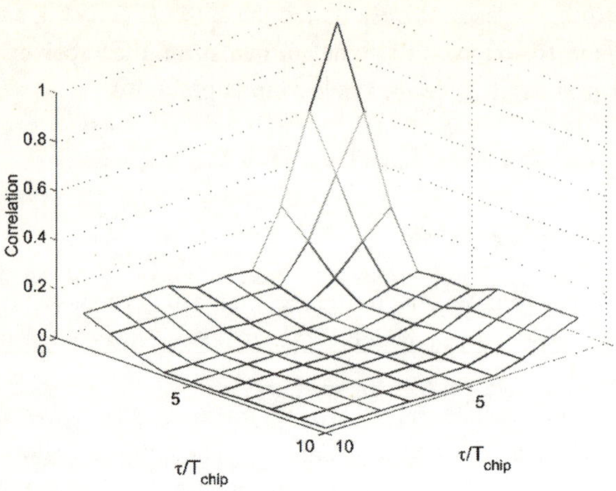

***Figure 6**. Long-term temporal signal covariance matrix.*

(16) $$\hat{\mathbf{Q}}^{(S)}_{\hat{k},\,\zeta}(\tau) = \sum_{k \neq \hat{k}}^{\hat{K}} \hat{\mathbf{R}}^{(S)}_{k,\,\zeta}(\tau) + \sigma_N^2 \mathbf{I}$$

the spatial covariance matrix including the total network interference for the $\hat{k}$*-th* user belonging to cell ζ. Following the description above and repeating the procedure for $\zeta = 1, \ldots, Z$ ($Z = 57$) cells, we can finally calculate the quality requirement at the BS antenna connector for each user in the network.

Introducing the processing gain *GP* as a function of the service of the $\hat{k}$*-th* user

(17) $$GP_{\hat{k},\,\zeta} = W / R_{\hat{k},\,\zeta}$$

with the modulation bandwidth W and the user data rate $R_{\hat{k},\,\zeta}$, we obtain the ratio of combined received energy per information bit to the effective noise power spectral density, N_{eff}, at a single point in time.

(18) $$\left.\frac{E_b}{N_{eff}}\right|_{\hat{k}} = \frac{RSCP_{\hat{k},\,\zeta}}{ISCP_{\hat{k},\,\zeta}} GP_{\hat{k},\,\zeta}.$$

$RSCP_{\hat{k},\,\zeta}$ and $ISCP_{\hat{k},\,\zeta}$ are *the Received Signal Code Power* and the *Interference Signal Code Power* (including noise) of the $\hat{k}$*-th* user in cell ζ *after* beam forming and rake combining.

For the computer simulation of the above model, we use a stationary *Monte-Carlo* approach [9, 22, 23]. The schematic flowchart of the system simulator is shown in Figure 7.

The main simulation parameters are given in Table I.

V. Results and discussion

This section is structured as follows. First we give the definition of the system capacity bound and the system load in the uplink of a CDMA system. Then we discuss the results of space-only processing and finally of space-time processing.

V.1.1. System load definition

The maximum load, i.e., the maximum capacity of a single cell in the UMTS FDD uplink, is defined as the number of users corresponding to a certain rise in multiple-access interference MAI over the thermal background noise [2]. This "noise rise" is the increase in the uplink noise at the BS antenna connector due to the uplink interference. It is expressed as a dB figure relative to the thermal noise floor (more precisely, the thermal noise including the man-made noise). The 95% confidence value of this noise rise is taken over all Monte Carlo runs and all monitored cells. Our chosen design target value for this figure will be 6 dB. This is equivalent to 75% of the pole-capacity loading of a CDMA system [2]. The pole capacity is the maximum theoretical capacity of an *isolated* CDMA cell [26].

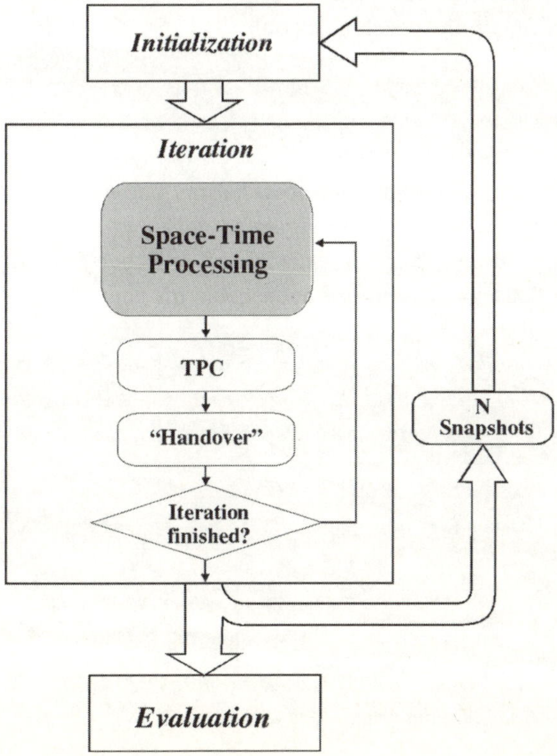

***Figure 7**.*
Simplified flowchart of the system simulator.

Figure 8 shows the relationship between uplink noise rise seen at the BS antenna and the system load η in %. The term η is defined as

(19) $$\eta = 1 - \frac{SIR_{loaded}}{SIR_{empty}}$$

where SIR_{loaded} and SIR_{empty} are the signal-to-interference-ratio at the BS antenna in a loaded cell (or system) and in an empty cell (or system), respectively [22].

V.1.2. Space-only results

We consider the flat fading propagation channel as described in Section 2. The individual multi-path components receive at $\tau = 0$ at the BS antenna.

We now make a comparative performance evaluation of simplified multi-service scenarios of the UMTS FDD uplink. We consider two systems, one equipped with single-element antennas and the other with four-element antenna arrays at all base stations within the network. Further we assume a service mix of speech users with a net data rate of 12.2 kbit/sec and 64 kbit/sec data users. As schematically explained in Figure 1, we expect that the gain of smart-antennas over the single-element antenna is higher, if the percentage of high-data-rate users in the network is high. Hence, we compare the noise rise of both systems for various service mixes, starting with speech users only.

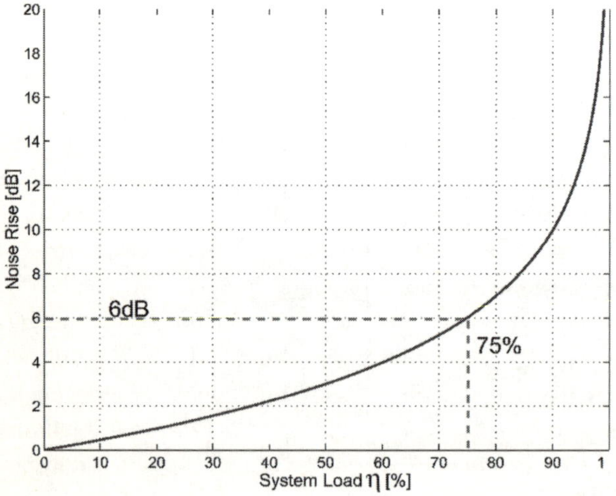

Figure 8. Theoretical interrelationship between the noise rise at the BS antenna and the system load η in %.

We can observe from Figure 9 that the shape of the noise rise in case of speech users only is a smooth one for both systems. Taking a closer look, e.g., at the noise rise at the 50 users/site-level, we find a value of 5.8 dB for the non-adaptive-antenna system, while the noise rise for the smart-antenna network is only about 2.8 dB. This 3 dB gain in the noise rise is in the range of what was expected for CDMA systems with speech users only [18, 19].

The noise rise is increasing when we add more users to the single-antenna-element system. In contrast, the noise in the smart-antenna network rises only slightly as a function of both the number of speech users and the increased number of 64 kbit/sec-data users.

CDMA systems are usually planned to operate at a system load of η = 50% (as defined in Equation 19, [22]). For this operation point, Figure 10 defines the plane of possible *business cases*, for both systems. A business case is the number of users for the different services in the system area. The dashed lines (with the marker for the single antenna case, without marker for the smart-antenna case) indicate the respective capacity maximum, i.e. for 50% of system loading. To the left of these lines, we can look for viable business cases. We can obtain an ensemble of possible user ratios in our system. One such user ratio (a "business case") with 54% data users will serve a total of 15 users (8 64 kbit/sec data, and 7 12.2 kbit/sec speech) in the conventional system, and about 48 users (26–64 kbit/sec data and 22 12.2 kbit/sec speech) by applying smart-antennas to the BSs. The actual capacity gain, of course, depends on individual realization of the network.

This directly leads to the question of the *effective throughput* as a function of the system load η and the percentage of high data rate users in the system.

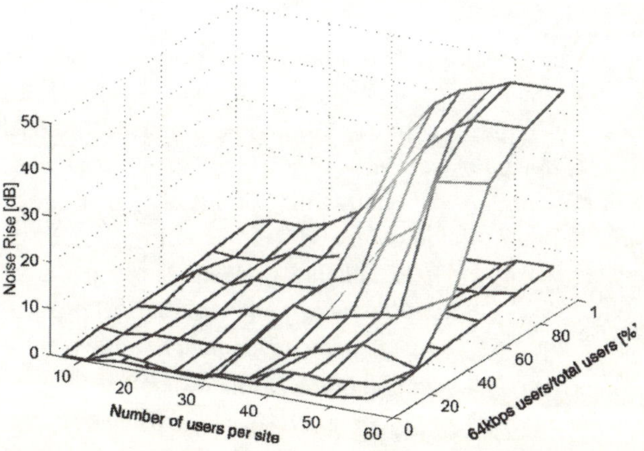

Figure 9. *Noise rise as a function of both the service mix of speech (12.2 kbit/sec) and 64 kbit/sec data users and the number of users per site. Two curves are shown. One for the noise rise for the system using 4-element antenna arrays, and a second one for the single element antenna. The latter completely overlays the first one.*

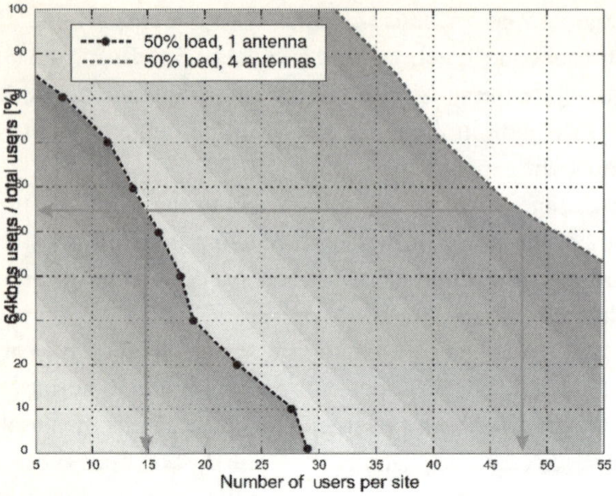

Figure 10. *Boundaries for a single-element antenna system (lower left area) and for a smart-antenna system (upper left area) as a function of the system load. The boundary is shown for a system load of* $\eta = 50\%$*. Arrows indicate a specific business case (54% data users).*

As indicated in Figure 11, the maximum effective throughput for individual business case realizations depends on the system load as well as on the service mix. For smart-antenna systems, the maximum throughput is steadily increasing with the number of high-data-rate users, but that is not the case for conventional antenna systems. This fact will be a main challenge for future UMTS operators and network planners. They have to optimize the network depending on the requirements of an increasing number of future data users, as well as on the system performance.

The plateaus, the kinks and bumps, are a consequence of our taking the 95% percentile value of CDFS.

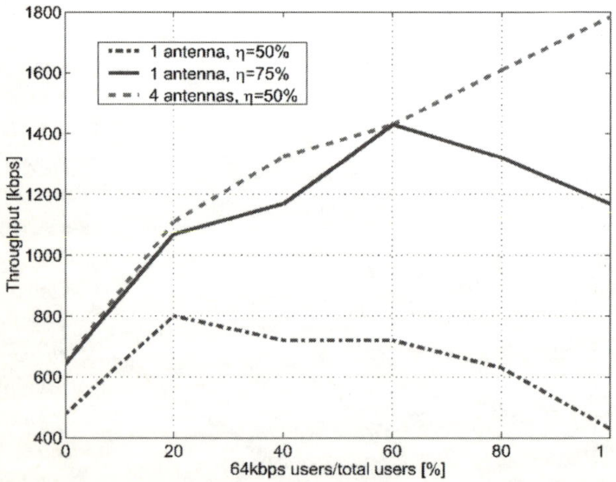

***Figure 11**. Effective throughput vs. the percentage of high-data-rate users in the system area.*

V.1.3. Space-time results

For the investigations on space-time performance on system-level we take the spatio-temporal channel model proposed in Section 2. Moreover we assume ideal channel estimation for the weight calculations.

We investigated a service mix of 40% speech users and 60% 64 kbit/sec data users. Figure 12 shows the results, comparing a single antenna and the 4-element antenna array case in a flat fading channel and the COST259 environment. For the temporal combiner we assumed a maximum of four selective rake fingers. Our investigation reveals a diversity gain due to rake finger assignment. Thus, the noise rise in the COST259 channel is consistently smaller than in the flat fading environment.

We also investigated 1, 4 and 6 antenna element uniform linear arrays with 2 and 4 rake fingers for the temporal combining as schematically shown in Figure 4. The results for a service mix of 40% speech users and 60% 64 kbit/sec data users are presented in Figure 13. We assumed a delay spread of 0.5 μsec [3] (about $2T_c$) and a number of 10 impinging multi-path components. In the case of 2 rake fingers only, the results are worse than in the flat fading environment, which corroborates the intuition that two rake fingers sum up only a portion of the power received [16, 29].

In general, the performance of rake receivers and the number of necessary rake fingers depend on the environment. In narrowband channels a low number of rake fingers is sufficient to sum up the signal energy. If we consider a urban channel with a high delay spread, more rake fingers are required [29]. This can be observed from Figure 13, where four rake fingers perform better than two in any antenna case.

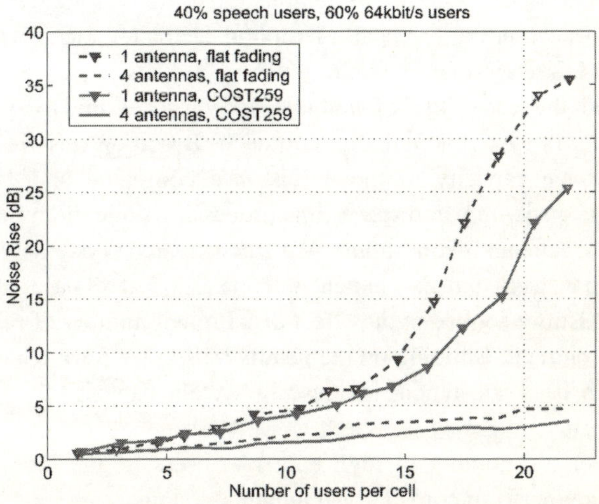

Figure 12. *Noise rise vs. the number of users per cell for a flat fading and the proposed spatio-temporal channel model.*

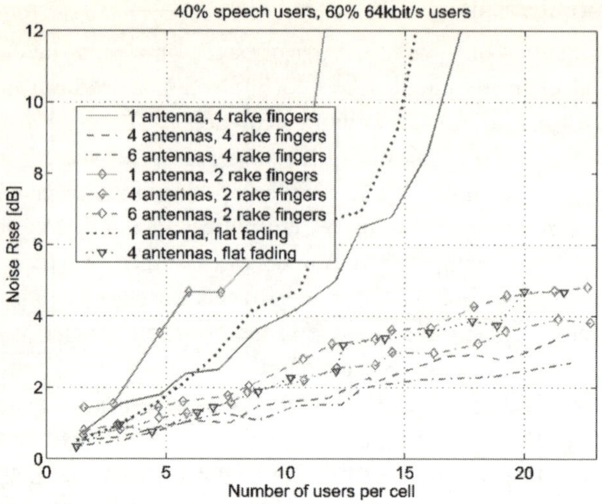

Figure 13*. Noise rise in the central cells. The* BS *are equipped with 1, 4 or 6 antenna elements and the temporal rake receiver combines 2 or 4 taps.*

Future UMTS base stations will provide at least four rake fingers per receiver branch. Thus, the flat fading assumption turns out to be a *lower* bound for the capacity estimation for both conventional and smart-antenna systems, respectively.

VI. Conclusions

The main focus of this study is the benefit of smart-antennas in the uplink of a multi-service UMTS FDD system. We demonstrated the importance of a realistic channel model accounting for spatio-temporal characteristics of mobile radio channels for a fair assessment of smart-antenna benefits.

We applied the capability of antenna array processing to our stationary Monte-Carlo UMTS system simulation approach. Based on this we investigated the mixed service capacity in the uplink. We compared and evaluated the performance of space-only and space-time processing under both flat fading and spatio-temporal channel conditions. We investigated systems with single-element antennas, four- and six-element antenna arrays at all base stations within the network. Also we looked at the effect of a limited number of rake fingers in an urban environment. Building on the results in Section 5 we conclude:

- There can be a substantial increase in system capacity by incorporating antenna arrays at the BS.
- The higher the number of high data rate users is, the more will smart-antennas outperform conventional antenna systems.

- The maximum effective throughput for individual business case realizations depends on the service mix. For smart-antenna systems the maximum throughput rises with the number of high-data-rate users in the system. This is not the case for single antenna networks.
- Contrary to GSM, the maximum throughput is a function of the system load. Therefore, network planning and optimization will be a main challenge for future UMTS operators.
- The flat fading assumption is a lower bound for the capacity estimation in urban scenarios.

Overall, we have shown a significant increase of mixed service capacity of UMTS uplink by smart-antennas and thus we strongly recommend their use at the BS.

Acknowledgement

The authors would like to thank Klaus Hugl, Werner Weichselberger and Thomas Baumgartner for fruitful discussions. They also acknowledge the useful editorial comments by Martin Töltsch and the reviewers of this paper.

REFERENCES

[1] UMTS-Forum, "The Future Mobile Market", Report n° 8, www.umts-forum.org, (March 1999).

[2] 3gpp, "RF System Scenarios – TS 25.942", http://www.3gpp.org, (April 2000).

[3] 3GPP, "Deployment aspects – TS 25.943", http://www.3gpp.org, (April 2000).

[4] CORREIA (L.H.), ed., Wireless Flexible Personalized Communications – COST 259: European Co-operation in Mobile Radio Research. *J. Wiley & Sons*, (March 2001).

[5] PEDERSEN (K.I.), MOGENSEN (P.E.), FLEURY (B.H.)," A stochastic model of the temporal and azimuthal dispersion seen at the base station in outdoor propagation environments", *IEEE Transactions on Vehicular Technology*, **49**, n° 2, pp. 437-447, (March 2000).

[6] BELLO (P.A.), "Characterization of Randomly Time-Variant Linear Channels", *IEEE Trans. on Communication Systems*, **CS-11**, pp. 360-393, (Dec. 1963).

[7] DELL'ANNA (M.), AGHVAMI (A.H.), "Performance of optimum and suboptimum combining at the antenna array of a W-CDMA system", *IEEE Journal on Selected Areas in Communications,* **17** n° 12, pp. 2123 –2137, (Dec. 1999).

[8] NAGUIB (A.F.), PAULRAJ (A.), "Performance of wireless CDMA with M-ary orthogonal modulation and cell site antenna arrays", *IEEE Journal on Selected Areas in Communications*, **14**, n° 9, pp. 1770 –1783, (Dec. 1996).

[9] NEUBAUER (T.), BAUMGARTNER (T.), BONEK (E.), "Necessary and Sufficient Network Size for Pole Capacity Estimation in UMTS FDD", *ECWT 2000,* Paris, France, pp. 131-134, (Oct. 2000).

[10] VITERBI (A.J.), VITERBI (A.M.) "Other-Cell Interference in Cellular Power-Controlled CDMA", *IEEE Transactions on Communications,* **42**, n° 2/3/4, pp. 1501-1504, (Feb. 1994).

[11] KOPSA (K.), WEINMANN (R.), BRAUN (V.), TANGEMANN (M.), "Performance of Multiple Antenna Combining Algorithms in the Uplink of UMTS/FDD", *Globecom* 2000-11-27.

[12] A. PAULRAJ (A.), PAPADIAS (C.B.), "Space-time processing for wireless communications", *IEEE Signal Processing Mag.*, **14**, pp. 49-83, (Nov. 1997).

[13] THOMSON (J.), GRANT (P.), MULGREW (B.), "Smart antenna arrays for CDMA systems", *IEEE Personal Commun. Mag.*, pp. 16-25, (Oct. 1996).

[14] WINTERS (J.H.), "Optimum Combining in Digital Mobile Radio with Co-channel Interference", *IEEE Journal on Selected Areas in Comm.*, **SAC-2**, n° 4, pp. 528-539, (1984).

[15] GERLACH (D.), PAULRAJ (A.), "Adaptive transmitting antenna methods for multipath environments ", *IEEE Global Telecommunications Conference,* GLOBECOM '94, **1**, pp. 425-429, (1994).

[16] MOLISCH (A.F.), "Wideband Wireless Digital Communications", *Prentice-Hall,* (2000).

[17] KUCHAR (A.), TAFERNER (M.), TANGEMANN (M.), HOEK (C.) "Field trial with a GSM/DCS1800 smart antenna base station" *Vehicular Technology Conference,* Amsterdam, Holland, **1**, pp. 42-46, (1999).

[18] LIBERTI (J.C.), RAPPAPORT (T.S.), "Analytical Results for Capacity Improvements in CDMA", *IEEE Trans. on VT*, **43**, n° 3, pp. 680-690, (Aug. 1994).

[19] NAGUIB (A.F.), PAULRAJ (A.), KAILATH (T.), "Capacity Improvement with Base-Station Antenna Arrays in Cellular CDMA", *IEEE Trans. on VT*, **43**, pp. 691-698, (Aug. 1994).

[20] HAGERMAN (B.), MAZUR (S.), "Adaptive Antennas in IS-136 Systems", *IEEE 48th Vehicular Technology Conference 1998*, Ottawa, Canada, pp. 2282-2286, (May 1998).

[21] BORMANN (R.), FRERICH (M.), AHRENS (F.), HENß (T.), DAM (H.), BERG (M.), ANDERSSON (S.), ALMQVIST (D.), "Functional Tests of adaptive antenna base stations for GSM", *3rd EPMCC 1999*, Paris, France, pp. 323-327, (March 1999).

[22] WACKER (A.), LAIHO-STEFFENS (J.), SIPILÄ (K.), JÄSBERG (M.), "Static simulator for studying wcdma radio network planning issues", *IEEE 49th Vehicular Technology Conference 1999,* VTC' 99 spring, Houston, USA, (May 1999).

[23] DEHGHAN (S.), LISTER (D.), OWEN (R.), JONES (P.), "W-CDMA capacity and planning issues", *IEE Electronics & Communication Engineering Journal*, pp. 101-118, (June 2000).

[24] Fuhl (J.), Molisch (A.F.), Bonek (E.), "Unified channel model for mobile radio systems with smart antennas", IEE *Proceedings – Radar, Sonar and Navigation*, **145**, n° 1, pp. 32- 41, (Feb. 1998).

[25] Riba (J.), Goldberg (J.), Vàzquez (G.), "Robust Beamforming for Interference Rejection in Mobile Communications", IEEE *Trans. on Signal Processing*, **45**, n° 1, pp. 271-275, (Jan. 1997).

[26] Lee (J.S.), Miller (L.E.), "CDMA Systems Engineering Handbook", *Artech House Inc.*, (1998).

[27] Brunner (C.), Haardt (M.), Nossek (J.A.), "On space-time RAKE receiver structures for WCDMA", *Conference Record of the Thirty-Third Asilomar Conference on Signals, Systems, and Computers*, **2**, pp. 1546-1551, (1999).

[28] Khalaj (B.H.), Paulraj (A.), Kailath (T.), "2D RAKE receivers for CDMA cellular systems", IEEE *Global Telecommunications Conference, GLOBECOM* '94, **1**, pp. 400-404, (1994).

[29] Win (M.Z.), Chrisikos (G.), Molisch (A.F.), Sollenberger (N.R.), "Selective Rake Diversity in Multipath Fading with Arbitrary Power Delay Profile", IEEE *Global Telecommunications Conference*, GLOBECOM 2000, San Francisco, (27 Nov.-1 Dec. 2000).

[30] Proakis (J.G.), "Digital Communications", *McGraw Hill Book* Comp. Inc., 3rd edition, New York, (1995).

[31] Pedersen (K.I.), "Antenna Arrays in Mobile Communications", Ph.D. Thesis, Center for PersonKommunikation, Aalborg University, Denmark, (Jan. 2000).

[32] Lancaster (P.), Tismenetsky (M.), "The Theory of Matrices", Second Edition, *Academic Press*, (1985).

[33] Cupo (R.L.), Golden (G.D.), Martin (C.C.), Sherman (K.L.), Sollenberger (N.R.), Winters (J.H.), Wolniansky (P.W.), "A four-element adaptive antenna array for IS-136 PCS base stations" IEEE *47th Vehicular Technology Conference*, **3**, (1997).

[34] Adachi (F.), Sawahashi (M.), Suda (H.), "W-CDMA: Performance evaluation and future enhancement", *3rd European Personal Mobile Communications Conference, EPMCC*, Paris, 9-11, pp. 429-434, (March 1999).

[35] 3GPP, "Vocabulary for 3gpp Specifications – TS 21.905", http://www.3gpp.org, (April 2000).

[36] Neubauer (T.), Jäger (H.), Fuhl (J.), Bonek (E.), "Measurement of the Background Noise Floor in the UMTS FDD Uplink Band", *4th European Personal Mobile Communications Conference, EPMCC*, Vienna, (Feb. 2001).

Chapter 6

Radio network planning process and methods for W-CDMA

J. Laiho and A. Wacker

Nokia Group, Helsinki, Finland

I. Introduction

As the launch of third generation technology approaches, operators are forming strategies for the deployment of their networks. These strategies must be supported by realistic business plans both in terms of future service demand estimates and the requirement for investment in network infrastructure. Evaluating the requirement for network infrastructure can be achieved using system dimensioning tools capable of assessing both the radio access and the core network components. Having found an attractive business opportunity, system deployment must be preceded by careful network planning. The network planning tool must be capable of accurately modelling the system behaviour when loaded with the expected traffic profile. The third generation cellular systems will offer services well beyond the capabilities of today's networks. The traffic profile, as well as the radio access technology itself, form the two most significant challenges when dimensioning and planning a W-CDMA based third generation system. The traffic profile describes the mixture of services being used by the population of users. There are also specific system functionalities which must be modelled including fast power control and soft handover. In order to accurately predict the radio coverage the system features associated with W-CDMA must be taken into account in the network modelling process. Especially the channel characterization, and interference control mechanisms in the case of any CDMA system must be considered. In W-CDMA network multiple services co-exist. Different services (voice, data) have different processing gains, E_b/N_0 performance and thus different receiver SNR requirements. In addition to these the W-CDMA coverage depends on the load characterization, hand over parameterization, and power control effects. In current second generation systems' coverage planning processes the base station sensitivity is constant and the coverage threshold is the same for each base station. In the case of W-CDMA the coverage threshold is dependent on the number of users and used bit rates in all cells, thus it is cell and service specific.

The W-CDMA planning process can be divided into three phases: initial planning (dimensioning), detailed radio network planning and network operation and optimization. Each of these phases requires additional support functions like propagation measurements, Key Performance Indicator definitions etc. In a cellular system where all the air interface connections operate on the same carrier the number of simultaneous users is directly influential on the receivers' noise floors. Therefore, in the case of UMTS the planning phases cannot be separated into coverage and capacity planning. In the case of the post-second generation systems data services start to play an important role. The variety of services requires the whole planning process to overcome a set of modifications. One of the modifications is related to the quality of service (QoS) requirements. So far it has been adequate to specify the speech coverage and blocking probability only. Also one has to consider more and more the indoor and in-car coverage probabilities. In the case of UMTS the problem is slightly more multi-dimensional. For each service the QoS targets have to be set and naturally also met. In practice this means that the tightest requirement shall determine the site density. In addition to the coverage probability the packet data QoS criteria are related to acceptable delays and throughput. Estimation of the delays in thc planning phase requires good knowledge of the user behaviour and understanding of the functions of the packet scheduler.

Common features between second and third generation coverage prediction also exist. In all the systems both of the links have to be analysed. In current systems the links tend to be in balance whereas in the case of third generation one of the links can be higher loaded than the other, and thus either one of the links could be limiting the cell capacity or coverage. The propagation calculation is basically the same for all standards, with the exception that different propagation models could be used. Another common feature is the interference analysis. In the case of W-CDMA this is needed for the loading and sensitivity analysis, in the case of TDMA/FDMA it is essential for frequency allocation. In order to fully utilize the W-CDMA capabilities, a thorough understanding of the W-CDMA air interface is needed from the physical layer to the network modelling, planning and performance optimization.

In this paper the pre-operational phase of the W-CDMA planning process, as depicted in Figure 1 in detail, is discussed. Section II concentrates on the initial planning issues. The W-CDMA link budget is introduced and it is demonstrated how different services and their QoS requirements impact on the site density estimate. In Section III a static radio network simulator is introduced. The

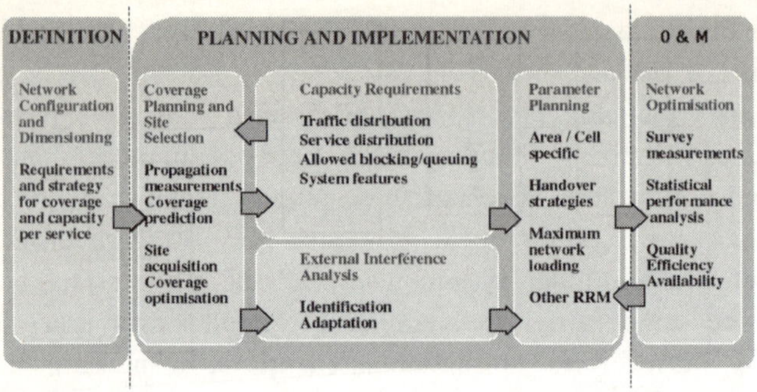

Figure 1. W-CDMA radio network planning process.

methodology required in the coverage and capacity estimation for W-CDMA is described for both uplink and downlink. The aim of Section IV is to demonstrate the accuracy of dimensioning: an example area is dimensioned and the average site distance is determined. The dimensioning result is compared with the outcome of a static network simulation. In Section V the analysis methods of the static simulator are verified. The verification was performed with a dynamic system simulator. The paper is concluded in Section VI.

II. Initial planning, system dimensioning

Initial planning (i.e. system dimensioning) provides the first and most rapid evaluation of the network element count as well as the associated capacity of those elements. This includes both the radio access network as well as the core network. This paper focuses upon the radio access part solely. The target of the initial planning phase is to estimate the required site density and site configurations for the area of interest. Initial planning activities include radio link budget (RLB) and coverage analysis, capacity estimation, and finally, estimation for the amount of base station hardware and sites, radio network controllers (RNC), equipment at different interfaces, and core network elements. The service distribution, traffic density, traffic growth estimates and QoS requirements are essential already in the initial planning phase. In the initial planning phase the quality is taken into account in terms of blocking and coverage probability. RLB calculation is done for each service, and the tightest requirement determines the maximum allowed isotropic path loss.

II.1 W-CDMA specific items in the radio link budget

In this section the W-CDMA uplink and downlink budgets are discussed. To estimate the maximum range of a cell a RLB calculation is needed. In the RLB the antenna gains, cable losses, diversity gains, fading margins, diversity gains etc. are taken into account. The output of the RLB calculation is the maximum allowed propagation path loss which in return determines the cell range and thus the number of sites needed. There are a few W-CDMA specific items in the link budget if compared with the current TDMA based radio access system like GSM. These include interference degradation margin, fast fading margin, transmit power increase and soft handover gain.

The interference degradation margin is a function of the cell loading. The more loading is allowed in the system, the larger interference margin is needed in uplink, and the smaller is the coverage area. The uplink loading can be derived as follows, for simplicity the derivation is performed with service activity $v = 1$.

To find out the required uplink transmitted and received signal power for a mobile station MS_k connected to a particular base station BS_n, the basic CDMA E_b/N_0 equation is used. The usual, slightly theoretical, assumption is that I_{oth}, the interference received from the MSs connected to the other cells, is directly proportional (proportionality constant i) to I_{own}, the interference received from the MSs connected to the same BS_n as the desired MS. Assume that the MS_k uses bit rate R_k, its E_b/N_0 requirement is ρ_k and the CDMA modulation bandwidth is W. Then the received power of the *k-th* mobile, p_k, at the base station it is connected to, must be at least such that

$$(1)\quad \frac{W}{R_k}\left(\frac{p_k}{I_{own}-p_k+I_{oth}+N}\right)=R_k\left(\frac{p_k}{I_{own}-p_k+iI_{oth}+N}\right)\geq \rho_k$$

$k = 1, \ldots, K_n$

where K_n is the number of MSs connected to BS_n, $N = N_0 W = N_f \kappa T_0 W$ is the noise power in the case of an empty cell, N_f is the receiver noise figure, κ is the Boltzmann constant and T_0 is the absolute temperature.

The inequalities in (1) are slightly optimistic because it is assumed that there is no interference from the own signal. In reality this is not exactly true in multi-path propagation conditions. Equation (1) is however still chosen to avoid taking multi-path interference into account twice, i.e., the E_b/N_0 requirements determined from link level simulations are presented so that N_0 means only noise and multi-path interference is visible in higher E_b/N_0 requirement to a certain BER performance. Solving the inequalities as equalities means solving for the minimum required received power (sensitivity), p_k:

$$p_k\left(1+\frac{\rho_k R_k}{W}\right)=\frac{\rho_k R_k}{W}\ (1+i)\, I_{own}+\frac{\rho_k R_k}{W}\ N \Rightarrow$$

$$(2)\qquad p_k=\frac{1}{1+\dfrac{W}{\rho_k R_k}}\ (1+i)\, I_{own}+\frac{1}{1+\dfrac{W}{\rho_k R_k}}\ N, \qquad k=1,\ldots,K$$

If the equations in (2) are summed over the mobile stations connected to BS_n then

$$\sum_k p_k=\left[\sum_k \frac{1}{1+\dfrac{W}{\rho_k R_k}}\ (1+i)\right]\sum_k p_k + \left[\sum_k \frac{1}{1+\dfrac{W}{\rho_k R_k}}\right] N \Rightarrow$$

$$(3)\qquad \sum_k p_k(1+i)=\frac{\left[\sum_k \dfrac{1}{1+\dfrac{W}{\rho_k R_k}}\ (1+i)\right]N}{1-\left[\sum_k \dfrac{1}{1+\dfrac{W}{\rho_k R_k}}\ (1+i)\right]}$$

since $I_{own}=\sum_k p_k$. If loading is defined as

$$(4)\qquad \eta=\sum_k \frac{1}{1+\dfrac{W}{\rho_k R_k}}\ (1+i)$$

this loading definition can be enhanced to include sectorization gain, ζ, and service activity, v:

$$(5)\qquad \eta=\sum_k \frac{1}{1+\dfrac{W}{\rho_k R_k v_k}}\ (1+i\cdot\zeta)$$

In [12] the uplink loading is estimated using equation

(6) $$\eta = \frac{1}{W} \cdot \sum_{j=1}^{m} R_j \cdot v_j \cdot \rho_j (1 + i)$$

where m is the number of services used. The difference between equations (5) and (6) are due to the fact that (6) does not include sectorization gain and that in the derivation starting from equation (1) the denominator is $I_{own} - p_k + iI_{own} + N$ rather than $I_{own} + iI_{own} + N$, which is only the case when $p_k << I_{own}$.

The downlink dimensioning is following the same logic as the uplink. For a selected cell range the total base station transmit power ought to be estimated. In this estimation the soft handover connections must be included. If the power is exceeded either the cell range ought to be limited, or number of users in a cell has to be reduced. For downlink the loading (η_{DL}) is estimated based on

(7) $$\eta_{DL} = \sum_{i=1}^{I} \left[\frac{\rho_i R_i v_i}{W} \left((1 - \alpha_i) + \sum_{n=1,\, n \neq m}^{N} \frac{Lp_{mi}}{Lp_{ni}} \right) \right]$$

where Lp_{mi} is the link loss from the serving BS m to MS i, Lp_{ni} is the link loss from another BS n, to MS i, ρ_i is the *transmit* E_b/N_0 requirement for the MS i, including the SHO combining gain and the average power raise caused by fast power control, N is the number of base stations, I is the number of connections in a sector and α_i is the orthogonality factor depending on multi-path conditions ($\alpha = 1$: fully orthogonal).

The term $\sum_{n=1,\, n \neq m}^{N} \frac{Lp_{mi}}{Lp_{ni}}$ defines the iDL.

Direct output of the downlink RLB is the single link power required by a user at the cell edge. The total base station power estimation must take into account multiple communication links with average ($\overline{LP}_{mi}$) distance from the serving base station. Furthermore, the multi-cell environment with orthogonalities a_i should be included in the modelling. More on the downlink loading and transmit power estimations can be found in [13]. In the RLB calculation in uplink direction the limiting factor is the mobile station transmit power, in downlink direction the limit is the total base station transmit power. When balancing the uplink and downlink service areas both links must be considered.

The interference degradation margin to be taken into account in the link budget due to a certain loading η (either in uplink or downlink) is

(8) $$L = 10 \cdot \log_{10} (1 - \eta)$$

Fast fading margin or power control headroom is another CDMA specific item in the RLB. Some margin is needed in the mobile station transmission power for maintaining adequate closed loop fast power control in unfortunate propagation conditions like the cell edge. This is applicable especially for pedestrian users where the E_b/N_0 to be maintained is more sensitive to the closed loop power control. The power control headroom has been studied more in [6] and [7]. Another impact of the fast power control is the transmit power increase. In the case of a slowly moving mobile station the power control is able to follow the fading channel and the average transmitted power increases. In its own cell this is needed to provide adequate quality for the connection and does not cause any harm, since increased transmit power is compensated by the fading channel. For neighbouring cells however this means additional interference. The transmit power increase *(TxPowerInc)* is used to reduce the reuse efficiency according to equation (9). In equation (4) *i* should be replaced with term *TxPowerInc* × *i* in case the mobile station transmit power increase is significant.

(9) $$F_r = \frac{1}{1 + TxPowerInc \times i}$$

Soft handover gain is discussed already in [3]. Handovers – soft or hard – provide gain against shadow fading by reducing the required fading margin. Due to the fact that the slow fading is partly uncorrelated between cells, and by making handovers the mobile can select a better communication link. Furthermore, soft handover (macro diversity) gives an additional gain against fast fading by reducing the required E_b/N_0 relative to a single radio link. The amount of gain is a function of mobile speed, diversity combining algorithm used in the receiver and channel delay profile. More about SHO gain can be found in Section II.3 .

II.2 Receiver sensitivity estimation

In the link budget the BS receiver noise density is estimating the noise level over one W-CDMA – carrier. The required receiver SNR contains the processing gain and the loss due to the loading. The loading used is the total loading due to different services on the carrier in question. The required signal power *(S)* depends on the SNR requirement, receiver noise figure and bandwidth.

(10) $$S = SNR \cdot N_0 \cdot W$$

where

(11) $$SNR = \rho \cdot \frac{R}{W \cdot (1 - \eta)}$$

(12) $$\text{and } N_0 \cdot W = N_f \cdot \kappa T_0 W$$

where N_f is the receiver noise figure, κ is the Boltzmann constant, T_0 is the absolute temperature and η is the loading. In some cases the basic noise/interference level is further corrected with the man-made noise term.

II.3. Shadowing margin and soft handover gain estimation

The next step is to estimate the maximum cell range and cell coverage area in different environments/regions (see the example in Table I). The RLB is estimating the maximum allowed isotropic path loss, from that value a slow fading margin, related to the coverage probability, has to be subtracted. When estimating the coverage probability the propagation model exponent and the standard deviation for the log-normal fading must be set. If the indoor case is considered the indoor loss is from 15 to 18 dB and the standard deviation for log-normal fading margin calculation is set to 10–12 dB. In the case of outdoor coverage typical standard deviation value is 7 to 8 dB. Typical propagation constants range from 2.5 to 4. Traditionally, the area coverage probability used in the RLB is for single cell case [1]. The required probability is 90 to 95% and typically this leads to 7–8 dB fading margin, depending on the propagation constant and standard deviation of the log-normal fading. Equation (13) estimates the area coverage probability for single cell case:

$$F_u = \frac{1}{2}\left[1 - erf(a) + \exp\left(\frac{1 - 2 \cdot a \cdot b}{b^2}\right) \cdot \left(1 - erf\,\frac{1 - a \cdot b}{b}\right)\right] \tag{13}$$

where $a = \dfrac{x_0 - P_r}{\sigma \cdot \sqrt{2}}$ and $b = 10 \cdot n \cdot 10 \cdot \log_{10} \dfrac{e}{\sigma \cdot \sqrt{2}}$,

P_r is the received level at cell edge, n is the propagation constant, x_0 is the average signal strength threshold and σ is the standard deviation of the field strength.

In real W-CDMA cellular networks the coverage areas of cells overlap and the mobile station is able to connect to more than just one serving cell. If more than one cell can be detected the location probability increases and is higher than determined for a single isolated cell. Analysis performed in [2] indicates that if the area location probability is reduced from 96% to 90% the number of base stations is reduced by 38%. This number indicates that the concept of multi-server location probability should be carefully considered. In reality the signals from two base stations are not completely uncorrelated, and thus the soft handover gain is slightly less than estimated in [2]. In [3] the theory of the multi-server case with correlated signals is introduced:

$$(14)\ P_{out} = \frac{1}{\sqrt{2\pi}} \int_{-\infty}^{\infty} e^{-x^2/2} \left[Q\left(\frac{\gamma_{SHO} - a \cdot \sigma \cdot x}{b \cdot \sigma} \right) \right]^2 dx$$

where P_{out} is the outage at the cell edge, γ_{SHO} is the fading margin in the case of soft handover, σ is the standard deviation of the field strength and for 50% correlation $a = b = 1/\sqrt{2}$. With the theory presented for example in [1] this probability at the cell edge can be converted to the area probability. In the W-CDMA link budget the SHO gain is needed. The gain consists of two parts: combining gain against fast fading and gain against slow fading. The gain against slow fading is dominating and it is specified as:

$$(15) \qquad G = \gamma_{single} - \gamma_{SHO}$$

If we assume 95% area probability, $n = 3.5$ and the standard deviation is 7 dB the gain will be 7.3 dB – 4 dB = 3.3 dB. If the standard deviation is larger and the probability requirement higher the gain will be more.

II.4. Cell range and cell coverage area estimation

Once the maximum allowed propagation loss in a cell is known, it is easy to apply any known propagation model for the cell range estimation. The propagation model should be chosen so that it optimumally describes the propagation conditions in the area. The restrictions of the model are related to the distance from the base station, the base station effective antenna height, the mobile antenna height and the frequency. One typical example for a macro cellular environment is Okumura-Hata. Equation (16) presents an example of a Okumura-Hata path loss model for an urban macro cell with base station antenna height of 25 m, mobile antenna height of 1.5 m and carrier frequency of 1950 MHz [4].

$$(16) \qquad Lp = 138.5 + 35.7 \cdot \log_{10}(r)$$

After choosing the cell range the coverage area can be calculated. The coverage area for one cell in hexagonal configuration can be estimated with:

$$(17) \qquad S = K \cdot r^2$$

where S is the coverage area, r is the maximum cell range and K is a constant, depending on the network topology. The number of sectors is typically from 1 to 3. In the case of W-CDMA reasonable values are up to 6 sectors. In the case of 6 sectors the estimation of the cell coverage area becomes problematic, since

Table I. Example of a W-CDMA RLB.

	UL		DL	
Transmitter power	125.00	a	1372.97	mW
	20.97	$b = 10 \log_{10}(a)$	31.38	dBm
Tx antenna gain	0.00	c	18.00	dBi
Cable/body loss	2.00	d	2.00	dB
Transmitter EIRP (incl. losses)	18.97	$e = b + c - d$	47.38	dBm
Thermal noise density	−174.00	f	− 174.00	dBm/Hz
Receiver noise figure	5.00	g	8.00	dB
Receiver noise density	−169.00	$h = f + g$	− 166.00	dBm/Hz
Receiver noise power	−103.13	$i = 10 \log_{10}(W) + h$	− 100.13	dBm
Interference margin	−3.01	j	− 10.09	dB
Required Ec/Io	−17.12	$k = 10 \log_{10}(E_b / N_o/(W/R)) - j$	− 7.71	dB
Required signal power [S]	−120.26	$l = i + k$	− 107.85	dBm
Rx antenna gain	18.00	m	0.00	dBi
Cable/body loss	2.00	n	2.00	dB
Coverage probability (outdoor requirement)	95.00		95.00	%
Coverage probability indoor (requirement)	0.00		0.00	%
Outdoor location probability (calculated)	85.62		85.62	%
Indoor location probability (calculated)	32.33		32.33	%
Limiting environment:	outdoor		outdoor	
Log normal fade constant outdoor	7.00		7.00	dB
Log normal fade constant indoor	12.00		12.00	dB
Propagation model exponent	3.50		3.50	
Log normal fade margin	− 7.27	o	− 7.27	dB
Handover gain (inc. any macro diversity combining gain at cell edge)	0.00	p	2.00	dB
Slow fade margin	− 7.27	$q = o + p$	− 5.27	dB
Indoor loss	0.00	r	0.00	dB
TPC headroom (fast fade margin)	0.00	s	0.00	dB
Allowed propagation loss	147.96	$t = e - l + m - n + q + r - s$	147.96	dB

a six-sectored site does not necessary resemble a hexagon. A proposal for the cell area calculation at this stage is that the equation for the omni case is used also in the case of 6 sectors and the larger area is due to a higher antenna gain. The more sectors are used the more careful soft handover overhead has to be analysed to provide an accurate estimate. In Table II some of the *K*-values are listed.

Table II. K-values for the site area calculation.

SITE CONFIG.	1 omni	2-sectored	3-sectored	6-sectored
K	2.6	1.3	1.95	2.6

II.5. Capacity and coverage analysis in the initial planning phase

Once the site coverage area is known the site configurations in terms of channel elements, sectors and carriers, and site density (cell range) are to be selected so that the supported traffic density can fulfil the requirements. An example dimensioning case can be seen in Section IV. The W-CDMA RLB is slightly more complex than the TDMA one. The cell range depends on the number of simultaneous users (number of channels/users in terms of interference margin, see equation (6)). Thus the coverage and capacity are connected and in the very beginning the operator should already have knowledge and vision of the subscriber distribution and growth since it has a direct impact on the coverage. Finding the correct configuration for the network so that the traffic requirements are met and the network cost minimized is not a simple task. The number of carriers, number of sectors, loading, number of users and the cell range all have an impact on the result.

III. Detailed planning process

III.1. Introduction to a static radio network planning simulator

In this study the simulator first introduced in [9] was used. It is of static nature and needs as inputs a digital map, the network layout and the traffic distribution in form of a discrete user map. Each of the users can have different terminal speed and uses a different service (bit rate, activity factor, which both can be different for uplink and downlink). Therefore each mobile station gets assigned an individual E_b/N_0 requirement imported from link level simulations. The simulator itself consists of basically three parts – initialization, combined uplink and downlink analysis and post processing phase. Following the initialization part, round after round in the main part of the tool, both the uplink and downlink for all mobile stations are analysed and after the iterations have fulfilled certain convergence criteria, in the final step, the results of the uplink and downlink analyses are post processed for various graphical and numerical outputs. On top of these results, for selected areas (which also can consist of the whole network)

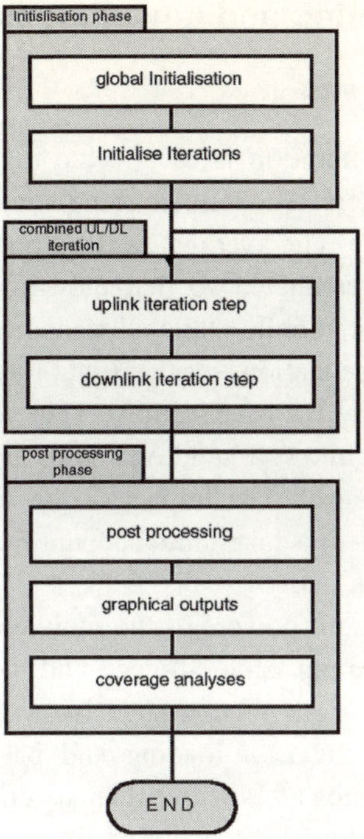

***Figure 2**. Static simulator overview.*

area coverage analyses for UL and DL dedicated channel as well as for common channels (common pilot CPICH, broadcast control channel BCCH, forward access and paging channel FACH and PCH on the P-CCPCH and S-CCPCH) can be performed. In case a second carrier is present in the network area, either used by the same operator or by another operator adjacent channel interference (ACI) can be taken into account. Only when the second carrier is assigned to the same operator, load can be shared according different strategies between the carriers (IF-HO).

III.2. Initialization phase

In the global initialization phase the network configuration is read in from parameter files for base stations, mobile stations and the network area. Some system parameters are set and propagation calculations are performed. In the following step, requirements coming from the link level simulations are assigned to base stations and mobile stations. After some initialization tasks for the iterative analysis – setting default transmit powers and network performance – the actual simulation can start.

III. 3. Combined uplink and downlink analysis

III.3.1. Uplink iteration step

The target in the uplink iteration is to allocate the mobile stations' transmit powers so that the (interference+noise)-levels and thus the base station sensitivity values converge. The average transmit powers of the mobile stations to each base station are estimated so that they fulfil the base station E_b/N_0 requirements. The average mobile stations' transmit powers are based on the sensitivity level of the base station, service (data rate) and speed of the mobile station and the link losses to the base stations. They are corrected by taking into account the activity factor, the soft handover gains and average power raise due to fast transmit power control. The impact of the uplink loading on the base station sensitivity (noise rise) is taken into account by adjusting it with $(1-\eta)$. η can be defined by equation (5).

After the average transmit powers of the mobiles have been estimated they are compared to the maximum value allowed and mobiles exceeding this limit are trying IF-HO if allowed or are put to outage. Now the interference analysis can be performed again and the new loading and base station sensitivities are calculated until their changes are smaller than specified thresholds. Also when the uplink loading of a cell exceeds specified limits, mobile stations are moved to another carrier if allowed (IF-HO). Otherwise they are put to outage.

III.3.2. Downlink iteration step

Similarly to the uplink the goal of the downlink iteration is to assign the base station transmit powers for each link (including SHO connections) a mobile station is having until all mobile stations receive their signal with the required carrier-to-interference-ratio, C/I, defined by equation (18).

$$targetCI = \frac{EbNo_{MS}}{W/R} \tag{18}$$

where $EbNo_{MS}$ is the received E_b/N_0 requirement of the MS depending on terminal speed and service. The actual received $(C/I)_m$ of MS m is calculated using MRC according to equation (19) by summing the C/I values of all links k, $k = 1, \ldots, K$ the mobile station m is having.

$$\left(\frac{C}{I}\right)_m = \sum_{k=1}^{K} \frac{p_{km}/Lp_{km}}{(1-\alpha_k)(p_k - v_m p_{km})/Lp_{km} + I_{oth,k} + N_m} \tag{19}$$

where P_k is the total transmit power of the base station to which link k is established, Lp_{km} is the link loss from the cell k to the mobile station m, α_k is the cell specific orthogonality factor, p_{km} is the power allocated to the link from base station k to mobile station m, $I_{oth,\,m}$ is the other cell interference and N_m is the background and receiver noise of MS m.

The initial transmit powers are adjusted iteratively according to the difference between the achieved and the targeted C/I value until convergence is achieved. The process requires iteration, since the C/I at each mobile station is dependent on all the powers allocated to the other mobile stations and it is not known a priori whether a link can be established or not. In case either certain link power limits or the total transmit power of a base station is exceeded mobile stations are performing IF-HO if allowed or taken out randomly from the network.

In a further step for each mobile station it is checked whether the received E_c/I_0 value is above a user defined threshold so that the mobile station can reliably measure the base station and synchronize to it. Also, here if the threshold given is exceeded, the mobile station tries IF-HO or is put to outage. A flow chart for the detailed iteration steps can be seen in Figure 3.

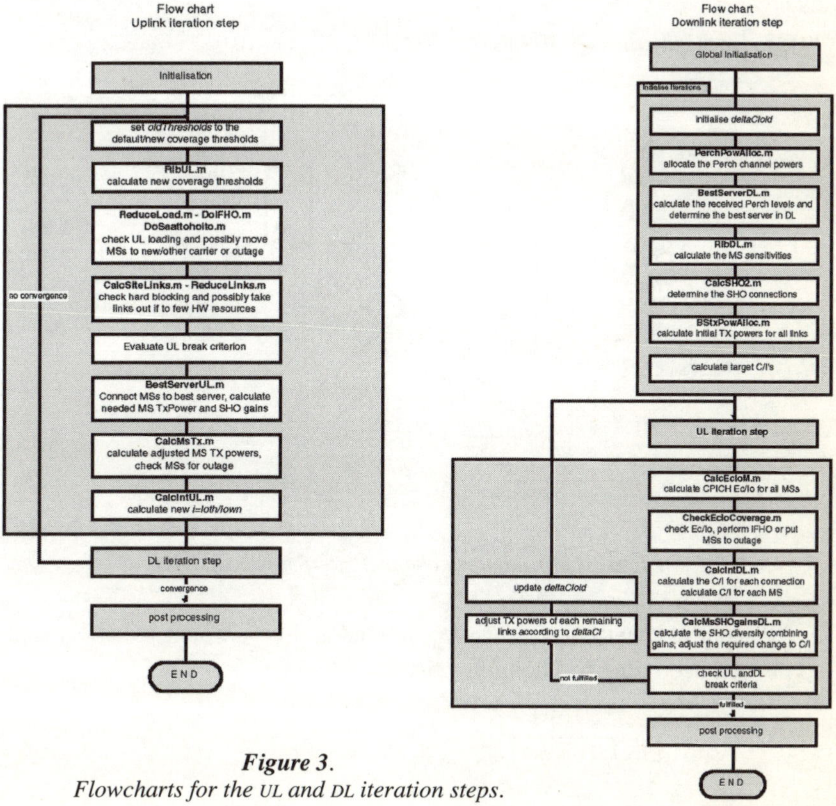

Figure 3.
Flowcharts for the UL and DL iteration steps.

III.3.3. Adjacent channel interference calculations

The adjacent channel interference influence due to the two possible carriers – either from own network or from a competitive operator's network in the same area – is taken into account by filtering this interference with a channel separation dependent filter. In both directions UL and DL the adjacent carrier filtering is implemented as a two-fold process. In UL, one filter for the mobile stations has been implemented indicating that it is out of band radiation (*acpFilterUL*). This filter is used to indicate how much power the mobile station is leaking into the other carrier's receiving band (*Adjacent Channel Leakage Ratio, ACLR*). For the base station in the uplink another filter (*aciFilterUL*) has been implemented. This filter is indicating the selectivity of the base station's receiver in multi-carrier situation, i.e. how big a portion of the adjacent channel power is received by the base station as adjacent channel interference power (*Adjacent Channel Protection, ACP*). Also this filter setting depends on the carrier separation. The adjacent channel interference situation in UL is depicted in Figure 4. In the simulations these two filters are combined to a single filter by equation (20):

(20) $$acFilterUL = -10 \cdot \log_{10}\left(10^{-\frac{aciFilterUL}{10}} + 10^{-\frac{acpFilterUL}{10}}\right)$$

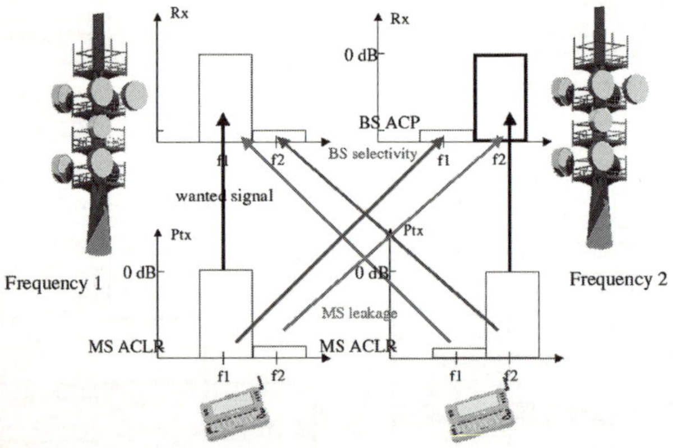

Figure 4. *UL adjacent channel interference situation.*

Adjacent channel interference (I_{ACI}) is taken into account when calculating the UL load according to Equation (21).

(21) $$\eta = \frac{I_{own} + I_{oth} + I_{ACI}}{I_{own} + I_{oth} + I_{ACI} + N}$$

Also in the downlink a similar type of filtering is introduced as in the uplink. One filter for the base stations has been implemented indicating the out of band radiation of the base station (*acpFilterDL*). This filter is used to indicate how much power the base station is leaking into the other carriers receiving band (*ACLR*). The filter setting depends on the separation between the carriers. For the mobile station another filter has been implemented (*aciFilterDL*). This filter is indicating the selectivity of the mobile station's receiver in a multi-carrier situation, i.e. how much of the adjacent channel interference power is received by the mobile station (*ACP*). Also this filter setting depends on the carrier separation. The adjacent channel interference situation in DL is depicted in Figure 5. In the simulations these two filters are combined to a single filter by equation (22):

$$acFilterDL = -10 \cdot \log_{10}\left(10^{-\frac{aciFilterDL}{10}} + 10^{-\frac{acpFilterDL}{10}}\right) \tag{22}$$

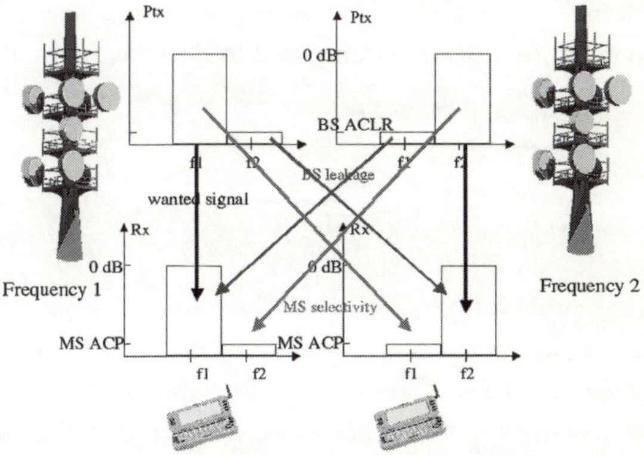

Figure 5. *DL adjacent channel interference situation.*

Adjacent channel interference (I_{ACI}) in DL is taken into account when calculating the C/I of a single link a MS is having according to equation (23).

$$\left(\frac{C}{I}\right)_m = \sum_{k=1}^{K} \frac{p_{km}/Lp_{km}}{(1-\alpha_k)(p_k - v_m p_{km})/Lp_{km} + I_{oth,\,k} + I_{ACI} + N_m} \tag{23}$$

where variables are as defined after equation (19), I_{oth} is interference from other cells on same carrier and I_{ACI} is adjacent channel interference.

III.4. Predicting the network coverage

In all estimations of area coverage probability (UL, DL, DCH, CPICH, BCH, FACH, PCH) it is assumed that the interference situation is stable. This means that a certain traffic distribution has been assumed and the detailed UL and DL iteration has been run. A test mobile is then run through all pixels of the interesting area and all other MSs which could be served are contributing to the interference. The test mobile is not influencing the interference situation, therefore the other to own cell interference ratio will not change and also the total transmit power of the serving BSs is constant.

III.4.1. UL DCH coverage

In UL direction it can now be estimated whether or not this additional mobile station using a certain bit rate and having a certain E_b/N_0 requirement could get the service in the chosen geographical location, meaning whether, if the maximum allowed transmit power of the test MS is enough to fulfil the E_b/N_0 requirement at the BS receiver. The transmit power needed for the MS is calculated with equation (24) and compared to the maximum allowed.

$$(24) \qquad P_{Tx,\,MS} = \frac{\mathrm{N}_0 \cdot Lp}{v \cdot (1-\eta) \cdot \left(1 + \dfrac{W}{R \cdot \rho \cdot v}\right)}$$

Area coverage probability is finally defined as the proportion of the chosen area where the additional MS really gets the wanted service under that stable interference situation. The weakness in this approach is that in reality an additional mobile station would change the interference situation, for example some other mobiles could go to outage but in the case of low bit rate services this effect can be neglected.

III.4.2. DL DCH coverage

In the downlink direction the coverage probability calculation is based on the transmit power limits per radio link. The main focus is to check, pixel-by-pixel, whether or not there is enough transmit power per link from base stations in the active set, if there would be a MS in the pixel, using a given service (bit rate) and having a given speed. Also here it is assumed that the total transmit powers of BSs do not change from what was left after UL/DL iteration, i.e. it could be assumed that the influence of the test MS is replaced by the closest original MS. In iteration however there may or may not have been a MS in the pixel. The coverage probability then again is defined as the number of pixels from the whole area where the service was possible.

III.4.3. CPICH coverage

In radio network planning the CPICH transmit power should be set as small as possible but ensuring that the best cell and neighbour cells can be measured and synchronized and the CPICH can be used sufficiently as the phase reference for all other DL physical channels. Typically this means that 5 – 10% of the total BS power is used for CPICH. For each pixel in the chosen area the E_c/I_0 in that pixel is calculated according to equation (25).

$$(25)\quad CPICHEcio = \frac{P_{CPICH}/Lp}{\sum_{i=1}^{numBSs} P_{TX,\,i}/Lp_i + I_{ACI} + N_0}$$

where P_{CPICH} is the CPICH power of the best server, Lp is the link loss to the best server, $P_{TX,\,i}$ is the total transmit power of BS i, Lp_i is the link loss to BS i, I_{ACI} is adjacent channel interference, N_0 is the thermal noise of the default MS and *numBSs* is the number of base stations in the network. The achieved E_c/I_0 is then compared to a user given threshold and the CPICH coverage defined as the ratio of pixels where the threshold is exceeded compared to all pixels. The weakness of this modelling for the CPICH E_c/I_0 coverage is that it is done only for the best server. In an operating network however, all neighbour cells must also be measured and therefore all neighbour cells' CPICH E_c/I_0 should be analysed, too. This could be overcome however by e.g. adding a threshold to the required CPICH E_c/I_0. Descriptions of other static simulators can be found for example in [18, 19].

IV. Example of dimensioning case and verification of dimensioning with static simulations

This section is contains an example of the case of the initial planning phase for a green field macrocellular operator. In this example only one network evolution phase is considered, in the real radio network planning case the traffic growth should be more carefully considered. In this work three cases have been analysed. The first case is 8 kbps speech only, mobile station speeds 3 (50% of MSs) and 120 km/h (50% of MSs). The second case is dimensioning speech and circuit switched data (64 kbps), speech mobiles moving 50 km/h, data terminals 3 km/h. The last case is 144 kbps data only, terminals moving with 20 km/h. More details related to the case can be found in [14]. The study consisted of two parts. In the first part the operator's macrocellular network is dimensioned, i.e. the cell range is estimated with given input parameters. This study is based on the assumption that the traffic and the QoS information are available from the operator. In the second phase the network is planned for the estimated site distance (1.5*R) and the W-CDMA analysis is performed for the radio network. As depicted in Figure 1 this case is concentrating on the first half of the radio network planning process: the network dimensioning, network configuration definition and coverage/capacity planning.

In the first phase the dimensioning task is to generate certain traffic and QoS requirements for the above mentioned three cases and to dimension the cases accordingly. The radio network was dimensioned for an urban area of 13×13 km^2. The propagation model used was Okumura-Hata with an area correction factor of 0 dB. The traffic and QoS requirements for all the three cases is shown in Table III.

Table III. *The traffic and QoS requirements and other dimensioning parameters.*

	SPEECH	SPEECH AND 64 KBPS CS DATA	144 KBPS PACKET DATA
Subscribers	60 000	60 000, 12 000	24 000
Average traffic/subscriber	0.060 Erl	0.040 Erl , 0.025 Erl	3 kbps
Blocking	2%	1%	-
Coverage probability	95%	95%/70%	80%
Data activity	50%	50%	100%
Assumed soft capacity	20%	10%	10%
SHO overhead	40%	40%	40%
Maximum uplink loading	60%	65%	65%
MS/BS transmit power	24/43 dBm	24/43 dBm	24/43 dBm
Common channel overhead	10%	10%	10%
Antenna gain MS/BS	0/17 dBi	0/17 dBi	0/17 dBi
Cable/body loss	3/0 dB	3/0 dB	3/0 dB
MS speed	3km/h 120 km/h	50 km/h, 3 km/h	20 km/h
Std for the log normal fading	7 dB	7 dB	7 dB
Peak to average ratio for total BS power calculation	4 dB	4 dB	4 dB
Average orthogonality	50%	50%	50%
i and iDL	55%	55%	55%
BS antenna height	30 m	35 m	35 m

In the dimensioning process both the capacity and the coverage have to be considered. The cell range is determined not only by the RLB, but also on the capacity requirements. From the selected cell range the coverage area for the site can be estimated by equation (17). Once the site coverage area is known the supported traffic density can be estimated. The selected site density must be such that the traffic density requirement and the coverage requirement in terms of coverage probability are met. In the following table the dimensioning results are collected. In the example case the cell range has been solely based on the uplink results. In the uplink direction the target loading as specified by equation (6) is limiting the number of channel elements per cell. As it can be seen already in the dimensioning results, the base station transmit power is actually limiting in all of the cases. In order to meet the capacity requirements of either cell range, the number of sectors or the number of carriers should be changed. In this case only one carrier was used.

In the second phase of the work the dimensioning results in terms of number of cells and the cell range were used in the corresponding static simulator runs. The items with **bold** font in Tables III and IV were used as inputs for the simulation. The site distance used in the static simulator has been estimated with D = 1.5*R where R is the cell range obtained from the RLB in dimensioning. In this study the network scenario was based on a regular grid, and the network area consisted only of urban area type. The static simulator

***Table IV**. Dimensioning results.*

	SPEECH	SPEECH AND 64 KBPS CS DATA	144 KBPS PACKET DATA
Selected cell range	2.08 km	1.98 km	1.52 km
Uplink loading	0.32 3 km/h 0.29 120 km/h	0.40 speech 0.18 data	0.65 data
Service limiting the cell range	Speech 3 km/h	64 kbps data	144 kbps data
Number of channel elements per cell required to meet the capacity based on UL	30 3 km/h 30 120 km/h	42 speech 6 data	855.6 kbps throughput ~ 6 data channels
Number of channel elements per cell supported by DL	21 3 km/h 19 120 km/h	27 speech 6 data	5 data
Required number of 3-sectored sites (based on UL)	20 (60 cells)	22 (67 cells)	37 (112 cells)

described in Section III was used in this study. The E_b/N_0 requirements and the traffic distribution used in the simulation were the same as were used during dimensioning. In the dimensioning phase only the antenna gain is used. In the simulations the antenna radiation pattern is also required and an antenna with 65° horizontal beamwidth was selected. The 65° antenna was selected because it is widely used already in 2G networks with three sectored configurations. Furthermore, according to [11] the 65° provides the optimum performance for the used network configuration. In all the simulations the antenna gain was always 17 dBi. The soft handover addition window was set to – 6 dB. In Figure 6 there is an example of the network scenario. The system features used in the simulations are from [10] and the ITU vehicular A channel [5] has been assumed for the channel delay profile.

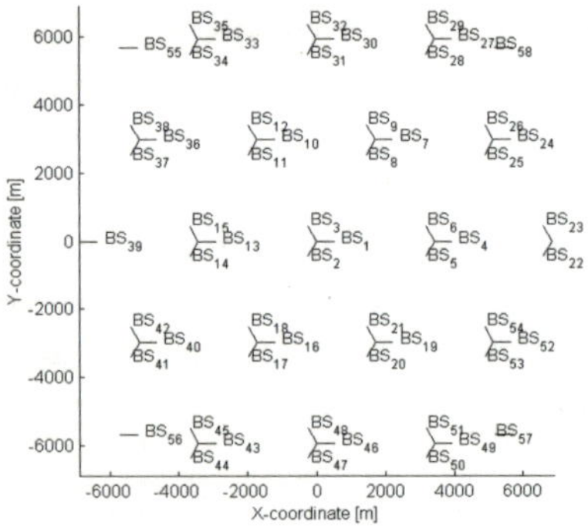

***Figure 6**. Example of a network scenario. Mixed traffic case.*

According to the simulations in the speech case the dimensioning is too pessimistic for the uplink capacity. Based on the simulations the network could support 72 speech users versus the 60 proposed by dimensioning. The main difference is because of the interference statistics. The other to own cell interference ratio has relative large standard deviation (0.27, minimum value only 0.11) and thus there are cells which can carry more traffic than equation (6) is proposing. With the help of the results in Table V it can be concluded that current dimensioning is correctly limiting the downlink capacity. According to the simulations with 65° antenna the downlink should serve 46 users, the dimensioning is proposing 40. The simulated coverage probability is slightly higher than the requirement, and thus the site distance in reality could be slightly increased.

In the mixed traffic case the uplink dimensioning result was proposing 42 channel elements for speech users and 6 for data users to meet the traffic requirement. The UL simulation result is slightly different from the dimensioning result. The reason for this is partly that it is very difficult to control the simulation run so that the ratio of the data users and the speech users would be exactly as required. The initial requirement was that the data users are 17% of the total users. After the static simulator run the percentage was 16%. In the case of UL there are on average 37 speech users instead of 42, but on the contrary there are 8 data users instead of 6. Therefore one can claim that the uplink dimensioning result is well in line with the simulation results. The power amplifier dimensioning for the DL (based on [13]) shows a clear downlink limitation. The uplink 48 users cannot be served in downlink direction with the assigned 20 W maximum base station transmit power. When the number of users in the dimensioning was reduced to 27 speech and 6 data users the dimensioning estimates the total TX power of 17 W including 10% common channel overhead. Values used for the power calculation were

iDL (= $\sum_{n=1,\, n \neq m}^{N} \frac{Lp_{mi}}{Lp_{ni}}$, see [13]) 55% and peak to average ratio for

Lp_{mi} of 4 dB. The downlink is limiting also according to the network simulations. According to the simulation downlink analysis the network can support 25 speech users and roughly 5 64 kbps data users simultaneously, the base station total transmit power being 19 to 20 W, depending on the selected scenario. In the mixed traffic case the dimensioning of the base station total transmit power follows extremely well the simulations. In terms of coverage probability the dimensioning is slightly pessimistic. According to the simulations the data coverage probability is always better than the required 70%. Since the data service is limiting the cell range the speech coverage probability is always better than the required 95%.

In the data case the dimensioning was proposing 855.6/144 = 5.94 simultaneous users (assuming 10% soft capacity, and iDL = 55%). This result is very close to the simulated case, but also in this case the uplink dimensioning is slightly pessimistic. In the downlink dimensioning with iDL = 55%, orthogonality 50% and 4 dB peak to average ratio the BS transmit power for the 5 users is roughly 19 W, including common channels (10% overhead) and 40% soft handover probability. A similar number is also achieved with the static simulations. In the data case the coverage probability is up to 8% higher than required. The difference in decibels of coverage probability of 80% and 88% is 2 to 3 dB, depending on the exact parameter values that have been used in the estimations, see [1]. In coverage limited case this has an impact on the required number of sites. All simulation results are collected in Table V.

In Figure 7 there is an example of the dominance and coverage analysis plot for speech services, test mobile speed 50 km/h. In the dimensioning the requirement used for speech was 95%.

Table V. Simulation results. Only the average results (averaged over all the cells) are included.

	SPEECH	SPEECH AND 64 KBPS CS DATA	144 KBPS PACKET DATA
Antenna 65°, site distance D = 1.5*R			
Number of users UL	35 3 km/h 36 120 km/h	37 speech 8 data	7
Number of users DL	23 3 km/h 23 120 km/h	25 speech 5 data	5
UL coverage probability	96.3% 3 km/h 98.8% 120 km/h	97.6% 74.4%	87.5%
Other to own cell interference	0.65	0.62	0.75
Loading	0.58	0.67	0.62
BS transmit power/W	19.34	18.95	19.10

For the mixed traffic (speech and 64 kbps) case the probabilities are 98% and 70% respectively. Since data services are limiting the range the actual coverage probability for speech was higher, i.e. 98%. Similar analysis for the RT data gave 68% coverage probability. In the dimensioning the used requirement was 70%.

The results presented here demonstrate that in macrocellular environment the downlink dimensioning is essential, due to the fact that with the assumptions of the macrocellular environment and Vehicular A channel conditions the network is downlink limited.

Uplink dimensioning alone gives too optimistic capacity predictions for the network. Furthermore, the modelling proposed for the dimensioning is verified with static network simulations. Comparison of the simulation results and the dimensioning results shows good agreement.

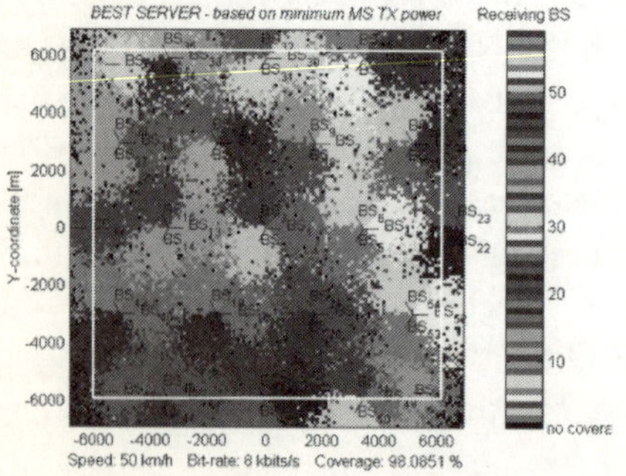

Figure 7*. Example result of the mixed traffic case. The coverage probability for speech services.*

V. Comparison of a static with a dynamic network simulator

The scope of this section is to demonstrate that the accuracy of the static simulator is adequate for radio network planning purposes.

V.1. Introduction to the dynamic simulator

Typically system level simulators operate with the resolution determined by the feature that changes interference situation most often. In W-CDMA the fast closed loop power control operating with 1.5 kHz frequency is the algorithm having the highest frequency, and therefore 1.5 kHz frequency is used in the system simulator used in the comparison. Conventionally, the information obtained from the link level tool is linked to the system simulation by using a so-called *average* value interface describing the BLER performance by average E_b/N_0 requirements. The average value interface is not accurate if there are fast changes in the interference due to, e.g. high bit rate packet users. This kind of approach suits well static snap-shot simulations, but cannot be used when simulating systems with fast power control and high bit rate packet data. With the dynamic simulator presented however, a so-called *actual* value interface (AVI) is used that provides accurate modelling of fast power and high bit rate packet data [16].

In the dynamic simulator the users are making calls and transmitting data according to the traffic models. The call generation process for real time services, such as speech and video, is made according to a Poisson process [15], [17]. For speech, voice activity and discontinuous transmission have to be considered. For circuit switched data services, the traffic model is a constant bit rate model, with 100% of activity.

The calculation of interference is an essential process of the system simulator. The better the interference modelling is, the more accurate results can be obtained. The total interference power $I_{bs(k)}$ received by a base station k is calculated as follows:

$$(26) \quad I_{bs(k)} = \sum_{\substack{n=1 \\ n \neq m}}^{N} \left[Lp_{n,k} \cdot \frac{\sum_{i=1}^{J} g_{i,n,k}}{\sum_{n=1}^{J} \hat{g}_{i,n,k}} \cdot p_{ms(n)} \right]$$

where N is the total number of active mobile stations in the system and m is index for the observed user. $Lp_{n,k}$ is path loss (attenuation due to distance and slow fading) between the base station k, and the mobile station n. $\sum g/\sum \hat{g}$ is the multi-path fading normalized to having long term average equal to one and J is the number of multi-path components. $p_{ms(n)}$ is the transmission power of the mobile n. After the interference calculations, the uplink signal-to-noise ratio SNR_{UL} can be calculated for the user m connected to the base station k as

$$SNR_{UL(m,\,k)} = \sum_{i=1}^{J} \frac{G \cdot p_{ms(m)} \cdot a_i^2}{I_{bs(k)} + N} \tag{27}$$

where G is the processing gain, a_i is amplitude attenuation of path i and J is the number of allocated RAKE fingers. In (27) it is assumed that the signals received are combined coherently with maximal ratio combining. In downlink the effect due to orthogonal codes has to be considered. Due to the multi-path propagation perfect orthogonality cannot be assumed. For optimal maximal ratio combining, the downlink signal-to-noise-ratio SNR_{DL} for a user m can be calculated as

$$SNR_{DL(m)} = \sum_{k=1}^{M} \left(\sum_{i=1}^{J_k} \frac{G \cdot p_{bs(m,\,k)} \cdot a_{k,\,j}^2}{I_{ms(m)} - P_{bs(k)}) \cdot a_{k,\,j}^2} \right) \tag{28}$$

where $I_{ms}(m)$ is the total interference power received by the mobile station m, M is number of base stations in the active set, $p_{bs(m,k)}$ is the transmitting power for the observed user from the base station k, $P_{bs}(k)$ is the total power transmitted from the base station k, $a_{k,i}$ is amplitude attenuation of the channel tap i and J_k is the number of allocated RAKE fingers from base station k.

In the dynamic simulator the following items were measured: bad quality calls, defined as calls having an average frame error rate FER exceeding a threshold (usually 5% for speech). The minimum call duration is set to 7 sec in order to increase the confidence of the averaging. Statistical data of these calls such as coordinates, start and end time and the call duration are recorded . Dropped calls, i.e. calls that have consecutive frame errors that exceed a threshold (usually 50 frame errors). Usually dropped calls are considered as severely poor quality calls. So bad quality and dropped calls can be taken as one measure whose percentage is referred to the number of started calls after the warm-up period. Power outage – for speech, this is taken from active terminals including those that are in DTX. Therefore it is slightly distorted due to the other half of the users that are in DTX. So the actual outage for terminals that are actively "talking" is higher, rough value is twice than that of the output. There is no discrepancy for data. E_b/N_0 targets are taken from all active terminals including those in SHO. So all factors regarding the channel and diversity are taken into account. Finally a SHO probability histogram of the number of branches per user was collected.

From the static simulator for all UL and DL connections the histogram of the transmit powers and their cumulative distribution function are taken. Moreover the p-th percentiles Q_p for 0, 50, 75, 90, 95 and 100% are extracted. Statistics showing the number and type of SHO connection were gathered. For the whole simulated area the estimated active set size (AS size) is collected (based on the CPICH levels.) In each simulated case also the UL loading level was stored.

In the final comparison the total traffic per cell, UL power distribution [dBm], DL total/link power distribution [dBm], SHO statistics, SHO areas, cell dominance areas, and not served mobiles (static) versus dropped and bad quality calls (dynamic) were of interest.

V.2. Comparison results

In this section some of the comparison results are collected. The main conclusion is that the cell level results (for example loading) are in good agreement with both the simulation methods. In Figure 8 the number of links is depicted cell by cell. It can be seen that the number of links per cell follows the same trend.

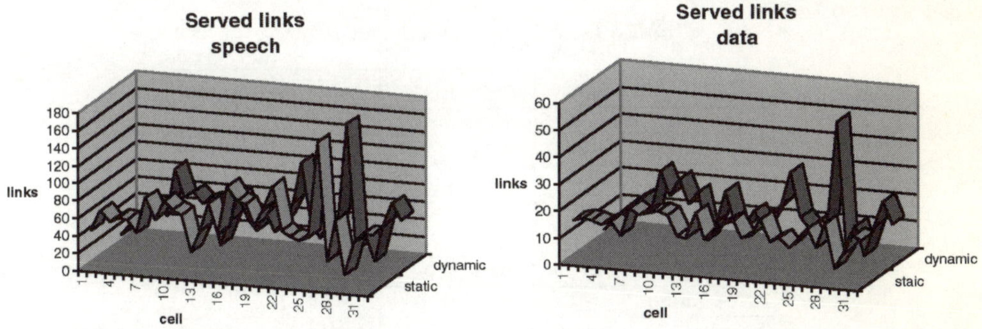

***Figure 8**. Number of links per cell for static simulator and dynamic simulator cell by cell.*

Figure 8 the difference of the cell dominance areas as seen by the two different simulators is depicted. It can be stated that the difference is minor. In 90–95% of all the pixels both simulators propose the same dominant cell.

The uplink power distribution statistics are collected for the speech and data case in Table VI. The maximum values do not differ significantly, but some of the percentile-values are well apart. This could indicate different power distribution shapes for the two simulators. This cannot be avoided due to the different nature of the simulators.

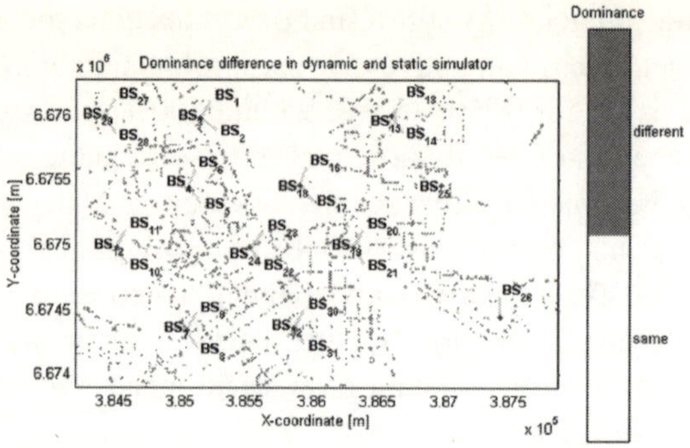

Figure 9. Dominance difference in the speech case.

In the downlink direction the transmit power statistics were collected. The comparison results are collected in Table VII. The difference in downlink direction has similar trend as in uplink.

The (downlink) link power distributions for the two simulators in speech case are depicted in Figure 10. The shapes of the distributions are close to each other. In the case of data the variance is larger.

***Table VI**. UL Power distribution difference.*

	MIN	Q50	Q75	Q90	Q95	MAX
Speech case						
Static s.[dBm]	−44	−10.38	−1.37	5.81	10.95	20.39
Dyn. s. [dBm]	−49	−14.5	−7	0	4.5	20
diff [dB]	−5	−4.12	−5.63	−5.81	−6.45	−0.39
Data case						
Static s.[dBm]	−41.79	−0.08	7.1	13.59	15.77	20.03
Dyn. s. [dBm]	−44	−3	6	15	19	20
diff [dB]	−2.21	−2.92	−1.1	1.41	3.23	−0.03

***Table VII**. DL Power distribution difference.*

	MIN	Q50	Q75	Q90	Q95	MAX
Speech case						
Static s.[dBm]	8.25	13.58	16.12	18.37	18.98	24.14
Dyn. s. [dBm]	−1	12.5	−7	20	21.5	24
diff [dB]	−9.25	−1.08	−0.38	−1.63	2.52	−0.14
Data case						
Static s.[dBm]	18.91	−24.01	24.8	25.57	25.83	26.29
Dyn. s. [dBm]	7	25	25.5	25.7	25.8	26
diff [dB]	−11.91	0.99	0.7	0.13	−0.03	−0.29

Another important result is the soft handover behaviour in the simulators. In Table VIII soft handover statistics from the data simulations have been collected.

In addition to the soft handover overhead the difference in active set sizes were investigated. These results are depicted in Figure 11.

In a radio network planning phase it is important to identify the outage areas of the network. In this study the outage prediction of the static radio network simulator was compared to the result from the dynamic simulator. The main conclusion of this case is that the problems have a tendency to be distributed roughly in the same locations for either static tool or dynamic one. The number of problematic calls cannot be directly compared to each other.

***Table VIII**. Handover comparison – 64 kbps CS data case.*

	HANDOVER TYPE			SHO
	1-way	2-way	3-way	OVERHEAD
Static sim. [%]	83.7	14.9	1.4	17.7
Dyn. sim. [%]	72	23	5	33

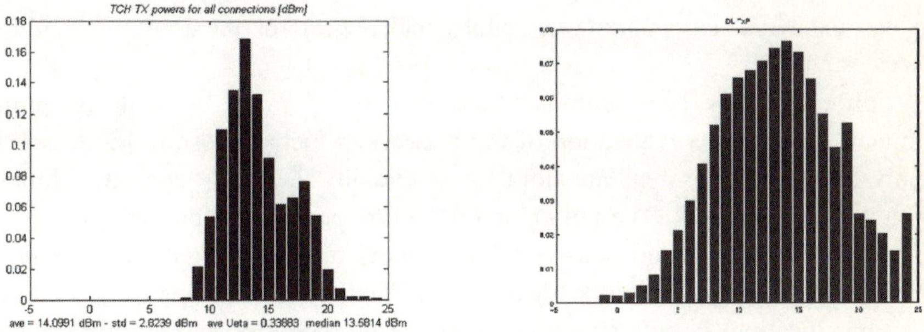

***Figure 10**. Link power distributions. Left figure for the static simulator (ave Ueta = average UL loading).*

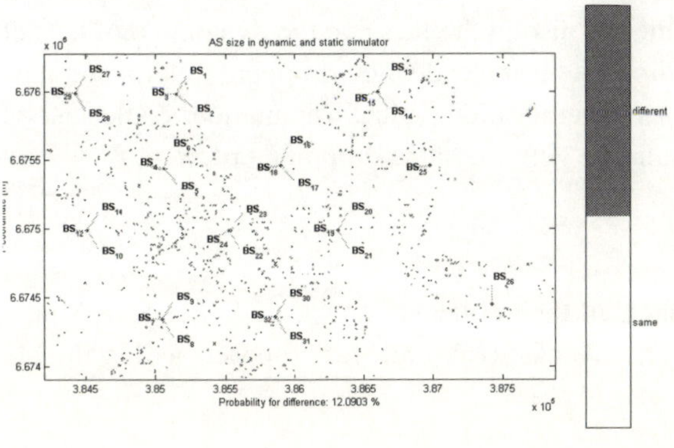

***Figure. 11**. AS size difference in speech case.*

VI. Conclusion

In this work the radio network planning process and methods for W-CDMA networks were introduced and verified. Accuracy of tools is essential to provide an operator with reasonable information of the required network topology and hardware requirements. In this work initial planning phase (dimensioning) methods were introduced. Furthermore the dimensioning results were compared with results proposed by static simulator. The comparison of the static simulation results and the dimensioning results show good agreement. In general it can be stated that the proposed dimensioning methods perform with reasonable accuracy in cases where the traffic distribution is such that the network can serve several simultaneous users in each cell. In the case of high bit rate services the network performance is strongly dependent on the location of the mobile users and the general interference situation, therefore it can be stated that also the proposed dimensioning modelling will perform with degraded accuracy. In the case of high bit rate services with low number of connections the proposed methods would underestimate the required BS power too much giving only a long-term average. One way to overcome this problem would be to use the worst case parameter values (cell edge) for the most demanding services.

Further, it has been demonstrated that static radio network planning simulator is giving a realization of the network, which is close to the network analysed by the fully dynamic simulator. Generally the tools result in a similar picture of the network. The uplink and downlink power distributions as well as the cell loading levels and supported links per cell are following the same trends. Problem zones in the network (dropped/bad quality calls, outage etc.) occur in the same locations in both analyses. Handover probabilities, active set sizes and dominance areas are almost the same in both tools. Nevertheless, a static tool is suitable for network planning. A dynamic simulator however is superior for benchmarking of radio resource management algorithms and for analysing the dynamic phenomena in the networks. The reason is not only in the computation complexity of the dynamic one but also in the fact that the dynamic tool is "not calibrated", i.e. that the call drop rate is dependent on dropping criteria and thus the number cannot be taken as exact absolute value. The number could just be compared with different simulation with the same dropping criterion.

Acknowledgements

The authors would like to thank all their colleagues, especially Mr. Kari Sipilä, Dr. Ari Hämäläinen and Dr. Tomáš Novosad for support and valuable contributions.

REFERENCES

[1] JAKES (W.C.), Microwave mobile communications, *IEEE PRESS*, p. 126.

[2] REUNANEN (J.), "Multiple Server Location Probability in GSM/DCS1800 Cellular Systems", Master of Science in Engineering Thesis, Helsinki University of Technology, (1997).

[3] VITERBI (A.J.), CDMA Principles of spread spectrum communication, *Addison-Wesley*, p. 198, (1995).

[4] HATA (M.), "Empirical Formula for Propagation Loss in Land Mobile Radio Services", *IEEE Transactions on Vehicular Technology*, **VT-29**, n° 3, (Aug. 1980).

[5] Guidelines for Evaluation of Radio Transmission Technologies for IMT-2000, Recommendation ITU-R M. 1225, (1997).

[6] SIPILÄ (K.), LAIHO-STEFFENS (J.), JÄSBERG (M.), WACKER (A.), "Modeling the Impact of the Fast Power Control on the WCDMA Uplink", *Vehicular Technology Conference*, Houston, (May 1999).

[7] SIPILÄ (K.), JÄSBERG (M.), LAIHO-STEFFENS (J.), WACKER (A.), "Soft Hand Over Gains in Fast Power Controlled WCDMA Uplink", *Vehicular Technology Conference*, Houston, (May 1999).

[8] TOSKALA (A.), HOLMA (H.), MUSZYNSKI (P.), "ETSI WCDMA for UMTS", *Proceedings of ISSSTA98*, South Africa, (Sept. 1998).

[9] WACKER (A.), LAIHO-STEFFENS (J.), SIPILÄ (K.), JÄSBERG (M.), "Static Simulator for Studying WCDMA Radio Network Planning Issues", *VTC99*, Houston.

[10] http://www.3gpp.org/.

[11] WACKER (A.), LAIHO-STEFFENS (J.), SIPILÄ (K.), HEISKA (K.), "The Impact of the Base Station Sectorisation on WCDMA Radio Network Performance", *VTC99* Amsterdam.

[12] A. SAMPATH (A.) *et al.*, "Erlang Capacity of a Power Controlled Integrated Voice and Data CDMA System", *IEEE 47th Vehicular Technology Conference Proceedings*, **3**, Phoenix, Arizona, pp. 1557-1561, (May 1997)

[13] SIPILÄ (K.), HONKASALO (Z.), LAIHO-STEFFENS (J.), WACKER (A.), "Estimation of Capacity and Required Transmission Power of WCDMA Downlink Based on a Downlink Pole Equation", in *Proc. of the IEEE VTC2000-Spring conf.*, Tokyo (May 2000).

[14] LAIHO-STEFFENS (J.), SIPILÄ (J), WACKER (A.), "Verification of 3G Radio Network Dimensioning Rules with Static Network Simulations", in *Proc. of the IEEE VTC2000-Spring conf.*, Tokyo, (May 2000).

[15] HÄMÄLÄINEN (S.), HOLMA (H.), SIPILÄ (K.), "Advanced WCDMA Radio Network Simulator", *Proceedings of PIMRC 1999*, Aalborg, Denmark, pp. 509-604, (Oct. 1997).

[16] Hämäläinen (S.), Slanina (P.), Hartman (M.), Lappeteläinen (A.), Holma (H.), Salonaho (O.), "A Novel Interface Between Link and System Level Simulations", *Proceedings of ACTS Summit 1997*, Aalborg, Denmark, pp. 509-604, (Oct. 1997).

[17] Brady (P.T.), "A Model for Generating On-Off Speech Patterns in Two-Way Conversation", *The Bell Systems Technical Journal,* **48**, n° 7, (Sept. 1969).

[18] Dehghan (S.), Lister (D.), Owen (R.), Jones (P.), "WCDMA capacity and planning issues", *IEE Electronics & Communication Engineering Journal,* pp 101-118, (June 2000).

[19] Labedz (G.), Love (R.), "A New Time-based Outage Criterion for the Forward and Reverse Links of DS-CDMA Cellular Systems", *VTC*98, Toronto, pp. 2182-2186.

Chapter 7

An open software-radio architecture supporting advanced 3G+ systems

C. Bonnet, G. Caire, P. Humblet, A. Nordio
D. Nussbaum and R. Knopp
Institut Eurécom, France

A. Enout
UDcast, France

G. Montalbano
Philips Semiconductors, France

T. Höhne
Nokia Research Center, Helsinki, Finland

B. Rimoldi
Mobile Communications Laboratory, Switzerland

I. Introduction and motivation

The presence of several different wireless communication standards and the wide variety of services provided by mobile communication operators has created the problem of providing universal seamless connection to customers with different service requirements at any point on the globe. Software Radio is an enabling technology for systems aiming at handling several different standards and different services and thus represents a solution to this problem. Generally, Software Radio is a very broad term encompassing several levels in the protocol stack (see e.g. [1, 2, 3] and references therein). Motivated by the worldwide activity around third generation (3G) mobile communication systems, Eurécom and EPFL have launched a joint project whose objective is to design and implement a real-time software radio communication platform to validate advanced mobile communication signal processing algorithms. The right to transmit has been granted for one 5 MHz UMTS channel in both France and Switzerland for experimental purposes.

The platform is characterized by the following major features:

- Flexibility, achievable by a software driven system.
- Duplex communication.
- Multiple antennas transmit and receive signal processing (i.e. joint spatio-temporal signal processing).

Flexibility remains a key word for a software-defined system. In our case it serves several purposes. For instance one may perform propagation channel measurements and transmitter characterization, evaluate the performance of different signal processing algorithms for both single user and multi-user systems under different operating conditions. *Duplex communication is* also necessary to allow higher-layer protocol testing and services, and to analyse more complex system aspects, such as multiple-access, power control, and optimize down-link signal processing from up-link measurements. In a second phase, the platform will allow *Multiple-antenna signal processing,* or more generally, spatio-temporal signal processing (also known as array processing) since this a very promising ensemble of techniques able to significantly increase the capacity of wireless communication systems.

The main focus of this paper is on the description of the basic platform architecture and signal processing to implement the essential physical layer level procedures of the UMTS standard operating in Time Division Duplex (UMTS/TDD) mode.

At the physical layer level, Software Radio generally requires the development of signal processing algorithms suited to implementation on a general purpose programmable processor (as opposed to analog or digital dedicated hardware). For this reason the position of Analog to Digital (A/D) and Digital to Analog (D/A) conversion must be moved as close as possible to the antennas. The goal is to perform operations like *channel selection* [4, 5], synchronization and detection in the all-digital domain, by using high performance Digital Signal Processors (DSPs). Special care must be dedicated to the transmission and reception front-end architecture. In fact, this should be independent of system-dependent parameters like the signal bandwidth and the symbol (or chip, in a CDMA system) rate. When designing software radio algorithms it is assumed that a multi-band Radio Frequency (RF) section takes care of translating the desired signal from a fixed Intermediate Frequency (IF) carrier to the required RF carrier in the transmitter, and vice-versa in the receiver. Then we are concerned with the efficient generation of an IF analog signal from a baseband digital signal (transmitter front-end) and with the reverse operation (receiver front-end). Once the end-to-end transmitter receiver architecture is defined the next step consists in synchronizing the receiver, estimating the channels associated with the users, and eventually detecting the transmitted symbols. All these operations are performed in real-time on our platform.

The adopted modulation scheme (BPSK and QPSK), the spread signal bandwidth (5 MHz), the spreading gain (DS-CDMA with allowed spreading gain of 1, 2, 4, 8, 16) and the frequency band (around 2.1 GHz) are the same as those defined in the UMTS/TDD specifications. The TDD mode has been chosen because under certain circumstances it allows the exploitation of the channel reciprocity between uplink and downlink in duplex communication, and also reduces the DSP computational load for a given bandwidth.

The solutions proposed in this paper can also be applied to a wide class of linearly-modulated digital signals (including most of today's and future mobile communications standards, like GSM, IS-54, IS-136, IS-95, and DECT (see e.g. [6] and references therein), UMTS (both FDD and TDD modes) [7], CDMA2000 (see e.g. [8, 9]) and EDGE (see e.g. [10] and references therein).

II. System architecture

In a first phase, the development of the platform has been based on a single antenna architecture for the Mobile Terminal (MT) as well as for the Base Station (BTS). The retained architecture permits to easy enhancement the capabilities of the platform without redesigning the essential hardware and software parts.

The BTS and the MT are based on the same hardware platform and differ only in the software implementation. The different hardware parts of the platform are highly partitioned in order to enable the use of as many standard cards and components as possible.

The hardware portion of the test-bed consists of 4 elements which are under software control, namely

1. a PCI bus based reconfigurable data acquisition card (DAQ) based on a Field Programmable Gate Array (FPGA)
2. an RF front-end
   - a single stage up/down-conversion from/to a 70 MHz intermediate frequency (IF) carrier with time-division-duplex (TDD) multiplexing
   - 1 high-speed 12-bit bandpass sampling A/D converter (f_{ADC} = 14.7456 MHz)
   - 1 high-speed 12-bit up-sampling D/A converter ($f_{DAC} = 8 \times f_{ADC}$ = 117.9648 MHz)
   - 8 slow D/A converters for controlling various amplifier gains on the RF card
   - control for various switches on RF card
3. a clock card for generating sampling clocks (fixed frequency) and local oscillators (programmable frequency)

We show a simplified block diagram of the entire system for a single antenna in Figure 1.

We have considered two software implementations, the first using a combination of commercially available embedded DSP cards and a common PC and the second using the DSP units (e.g. MMX) of a standard PC under an operating system proving hard real-time support (e.g. RTLinux). Both run in real-time and are compatible over the air.

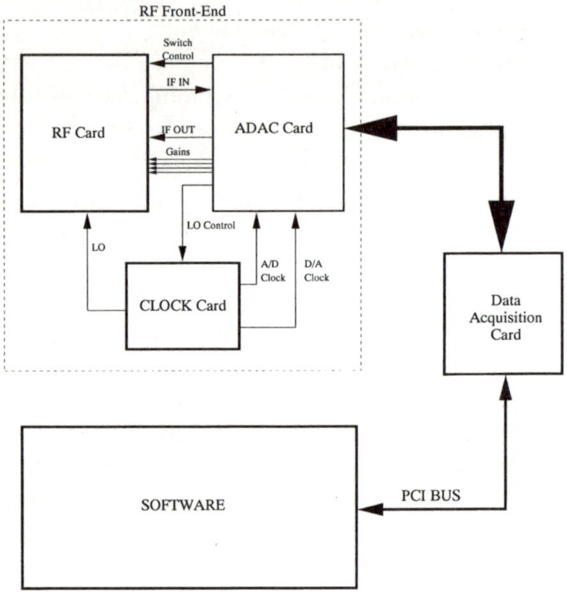

Figure 1*. System architecture.*

II.1. RF front-end

A simplified block diagram of our RF front-end is shown in Figure 2. It was designed in conjunction with STMicroelectronics in Geneva, Switzerland. On reception (Rx), the RF signal is filtered, amplified and down-converted to a 70 MHz intermediate frequency (IF). The local oscillator frequency is digitally tunable in steps of 500 Hz. The IF signal is amplified by a digitally tunable gain control and directly sampled at 14.7456 MSamp/sec. The samples are transferred via a ribbon cable using high-speed line drivers to the *data acquisition unit* (DAQ). The mode of the transceiver (i.e. transmission, reception, calibration) is fully controllable from the software portion of the platform.

On the transmission end (Tx), samples from the DAQ arrive at a rate of 14.7456 MSamp/sec and drive a hardware up-sampling circuit and a high-speed D/A converter (117.9648 MSamp/sec) to directly synthesize the 70

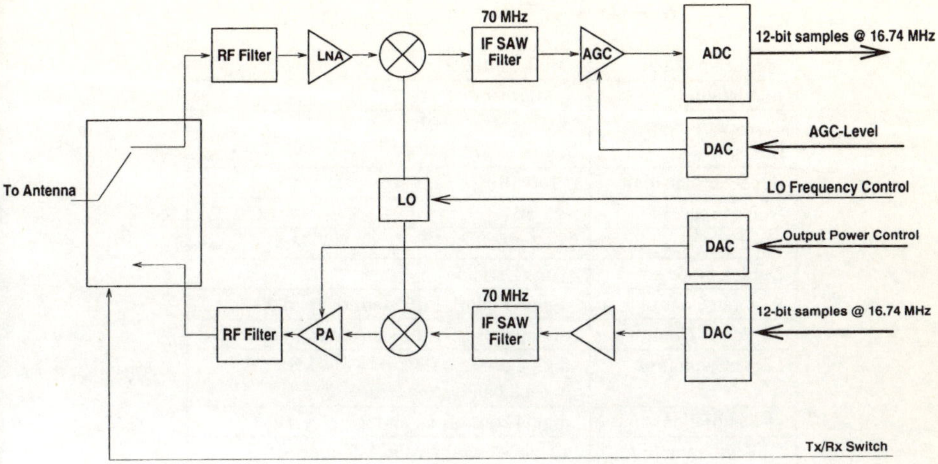

Figure 2. *RF front-end.*

MHz IF signal. This procedure is described in more detail in Section II.1.2. This signal is then amplified, up-converted to RF, filtered and amplified by a variable-gain power amplifier.

Special low-speed lines control the gains of the Rx variable-gain amplifier and Tx Power Amplifier, LO frequency, as well as the antenna switch. Although not included in Figure 2, automatic wide-band calibration capabilities (for both Tx and Rx) are included in the design, which are required for multiantenna systems (see e.g. [11]). These will be used in a later stage of the project where the architecture will be extended to implement multiple antenna transceivers. The basic characteristics of the RF front-end are summarized in Table I.

II.1.1. Passband A/D conversion

Once the RF signal at the antenna has been down-converted to IF, it is sampled by an A/D converter at a certain rate f_{ADC} (Figure 9). Calling $r_{\mathrm{IF}}(t)$ the received IF analog signal and choosing $f_{ADC} \geq 2W$ according to [4], because of the periodicity of the discrete-time signal spectrum, the resulting real sampled signal $r[n] = r_{IF}(n/f_{ADC})$ is pass-band with a spectrum replica centred at $f_{ADC}/4$ (although f_{IF} and f_{ADC} at the receiver can be different from f_{IF} and f_{ADC} at the transmitter, for simplicity we use the same notation).

Notice that here in order to avoid signal re-sampling we suppose the rate f_{ADC} to be a multiple integer of the chip rate (i.e. $f_{ADC} = N_c f_c$ where in our implementation we set $N_c = 4$).

Although a base-band version of the received signal can be obtained by multiplying $r[n]$ by $(-\mathrm{j})^n$ and then by low pass filtering, we will show further how the channel estimation and the data detection processes can be performed at pass-band.

Table I. RF *front-end characteristics.*

Frequency Band	2100-2170 MHz
Bandwidth	5 MHz (initially)
Transmit Power (per antenna)	1 W
Receiver Sensitivity	–100 dBm
Noise Figure	< 5 dB
Input IP3	> –7 dBm
Duplex mode	Time Division
Rx Gain Control	digital tuning, 1 dB steps over 40 dB
Tx Power Control	digital tuning, 1 dB steps over 80 dB range
Local Oscillator	digital tuning, steps of a few kHz in each band
RF Calibration	digital control, Tx and Rx
Direct IF sampling (70 MHz) on Rx	12-bit A/D @ 14.7456 MSamp/sec
Direct IF sampling (70 MHz) on Tx	12-bit D/A @ 117.9648 MSamp/sec
Digital Interface	High speed: Low-Voltage Differential Signaling (LVDS) Low speed: 3.3V CMOS line drivers

II.1.2. Passband D/A conversion

D/A converters have an impulse response $P_{DAC}(t)$ that can be approximated as a rectangular pulse of duration $1/f_{DAC}$ with frequency response of the form of $sinc(f/f_{DAC})$. If the discrete-time input of the D/A converter is the pass-band signal $x'[n]$, then the spectrum of the output signal is given by

$$Y(f) = \sum_i X'(f - if_{DAC})\mathrm{sinc}(f/f_{DAC}) \tag{1}$$

where $X'(f)$ is the discrete-time Fourier transform of $x'[n]$, defined by

$$X'(f) = \sum_n x'[n]e^{j2\pi nf/f_{DAC}}$$

Hence, the spectrum replica located at frequency f_{IF} is attenuated and distorted by the D/A impulse response as shown in Figure 8(a). A way to reduce the attenuation consists of using a D/A converter working at rate $f_d = L_{D/A}f_{DAC}$, where $L_{D/A}$ is a suitable integer, and up-sampling $x'[n]$ by the factor $L_{D/A}$. By choosing $L_{D/A}$ such that $f_d >> f_{IF}$, the spectrum replica around f_{IF} falls inside the first lobe of the D/A frequency response since its first zero is located at f_d (see Figure 8(b)). Moreover it is possible to compensate in part the distortion of the D/A converter by introducing a pass-band FIR filter between the up-sampler and the D/A converter. The filter must be designed in order to enhance the spectrum replica at IF while attenuating the other replicas.

Low complexity implementation. Denote by $x''[n]$ the up-sampled version of $x'[n]$, given by

$$x''[n] = \begin{cases} x'[k] & \text{for } n = L_{D/A}k \\ 0 & \text{otherwise} \end{cases} \quad (2)$$

and by $h_{D/A}[n]$ the filter impulse response. The filter output is given by

$$x'''[n] = \sum_{m} h_{D/A}[m]x''[N - M] \quad (3)$$

If the filter impulse response has length $L_{D/A}$, there is only a single non-zero term in the sum in the right-hand side of the above equation. Then, $x'''[n] = h_{D/A}[m]x'[k]$, where $k = [n/L_{D/A}]$ and $m = n$ modulo $L_{D/A}$. Therefore, the computational cost of the filtering operation consists of one real product per output sample at rate f_d. After the D/A conversion, the continuous-time signal $y(t)$ is eventually up-converted to the RF carrier and sent to the antenna.

II.2. Data acquisition card

The data acquisition card is a PCI bus-mastering device permitting high-speed full duplex parallel transfer of digital data from an external device. It contains the necessary glue logic which connects the input (A/D) and output (D/A) sample streams as well as some control signals to the main CPU/DSP. A PCI architecture was adopted since it is the most general purpose bus architecture and is used on most standard PCs, as well as DSP systems. It consists of two components, namely a powerful Xilinx Field Programmable Gate Array (FPGA) XCV300 [12] and a bus-mastering PCI controller PLX9080 [13]. The format is a *PCI Mezzanine Card* (PMC) to allow for integration into both embedded DSP architectures and ordinary PCs. A simplified overview of the DAQ is shown in Figure 3.

Firmware on the FPGA for formatting and transferring data via the DMA engines of the PLX9080 to the host (DSP, CPU) memory has been developed in VHDL *(Very High Speed Integrated Circuit Hardware Description Language)*. After an initial configuration phase, transfers are continuous and completely transparent to the host, who just "sees" a circular buffer containing samples acquired from or to be transferred to the external RF front-end.

The basic components of the DAQ are the following:

1. Line Drivers/Receivers for transfer from external devices via ribbon-cables.
2. A reconfigurable FPGA-based 16-bit bi-directional interface external device (up to 30 Msamp/sec full-duplex).

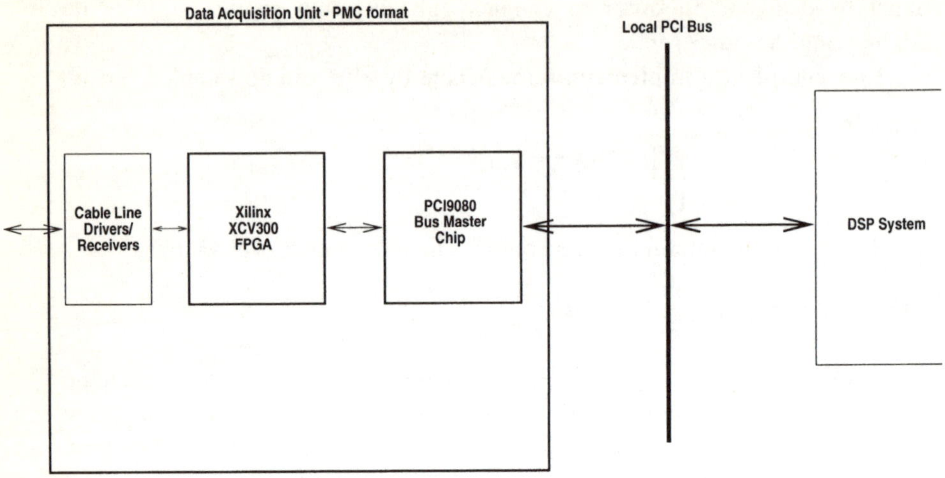

Figure 3. *Data acquisition unit.*

3. A PCI bus-mastering controller for direct transfer of samples to/from memory (DMA) on PCIbased signal processing units (e.g. DSP cards, high-performance PCs, workstations (SPARC, PowerPC, Alpha), embedded processor cards, etc.).
4. A *Processor Mezzanine Card (*PMC*)* form-factor for maximum flexibility. The FPGA is programmed via the PCI bus by serial download.

II.3. Texas Instruments TMSC6201 implementation

The embedded DSP architecture is based on a commercially available dual-DSP card (Spectrum Signal Processing Daytona [21]). The basic architecture is shown in Figure 4 and is centred around 2 Texas Instruments TMSC6201 fixed-point DSPs. These DSPs are capable of providing a maximum of 1600 MIPS each. Our DAQ is placed on the local PCI bus of the DSP board and transfers samples to/from both of the memory buffers on the DSPs buses. These buffers are used as temporary storage as the internal (fast) memory of the DSP is rather small. The DSP DMA engines take care of automatically transferring data to their internal memory concurrently with the signal processing functions. One DSP is used exclusively for transmission functions and the other for reception.

The DSPs are used for the front-end processing as described in the previous sections. Symbol rate data is transferred via the PCI bus to the PC which hosts the DSP card. This data is processed by the Pentium and handles tasks such as Viterbi decoding, carrier frequency offset compensation, higher layer protocol stacks, etc.

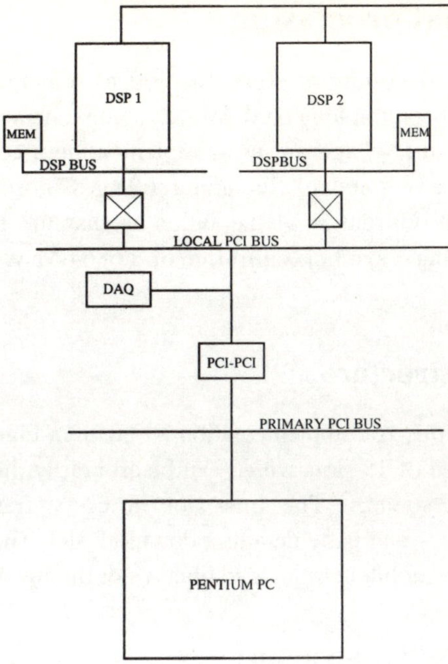

***Figure 4**. Embedded DSP architecture.*

II.4. RT Linux-based PC implementation

The second implementation does not rely on embedded DSPs. Here the DAQ is placed on the master PCI bus of a workstation, possibly multi-processor, running the hard real-time extension to the Linux operating system, RTLinux [22]. The software radio runs in kernel space and is integrated into the IPv4 (Internet Protocol) subsystem of Linux as a network device.

This x86 implementation makes use of the MMX (multi-media extensions) SIMD (single instruction multiple data) instructions for obtaining maximum processor efficiency for intensive DSP computations. All DSP routines use fixed-point arithmetic and are written in C with embedded assembly macros for time-critical code sections. We typically make use of:

1. MMX packed 16-bit arithmetic (multiply, add, MAC, interleaving, etc.);
2. loop unrolling;
3. software pipelining.

Due to the high-level software structure, this should be portable to other process or architectures (e.g. PowerPC, Alpha, etc.). For the same reason, it is easily portable to large-scale SMP (symmetric multi-processing) platforms which could be useful for advanced base-station implementations.

III. Digital signal processing

This section gives an overview of some theoretical principles on which the platform software has been implemented. We have implemented a variant of the UMTS-TDD 3GPP standard [14] and are now implementing a complete subset of layers 1 and 2 of the true standard (including 1.28 MChips/sec version). The main difference of the current implementation is that the hardware portion provides a clock yielding a symbol (chip) rate of 3.6864 Msymbols/sec and not 3.84 Msymbols/sec.

III.1. Frame/slot structure

The frame structure of this TDD implementation is shown in Figure 5. We see that each frame is composed of 15 slots which can be arbitrarily distributed between uplink and downlink streams. The first slot in every frame contains the synchronization sequence and is by default a downlink slot. The synchronization sequence is used by the mobile terminals to obtain slot timing synchronization.

III.2. Basic transmitter structure

The implementation of the transmitter allows for the generation a composite signal containing up to 8 variable-rate, variable power data streams per slot. It is shown in Figure 6. The rates of the different streams is controlled by OVSF spreading sequences $\phi_i[n]$ and the amplitudes by A_i. The choice of $\phi_i[n]$ dictates the spreading factor L_i, which ranges from 2^n, $n = 0, \ldots, 4$.

Two possible midambles, $m[k]$, can be inserted having lengths of either 256 or 512 chips. These are superposition of either 3 or 8 cyclic shifts of a training sequence with a periodic extension. This structure allows for efficient channel estimation techniques based on the *Fast Fourier Transform (FFT)*.

The beginning of synchronization slots (only BS) contains a primary synchronization sequence $s_p[k]$ of length 256 chips superimposed on the data.

The composite signal is filtered by a 12-tap root-raised cosine FIR filter, $p[k]$, which simultaneously up-converts the signal to a carrier frequency of $\pi/2$. To this end, we choose the sampling rate f_{DAC} according to the classical expression

(4) $$f_{DAC} = \frac{f_{IF}}{\ell \pm 1/4} \text{ for a positive integer } \ell$$

Then, we generate the discrete-time real signal

(5) $$x'[n] = \mathrm{Re}\{x[n]e^{j2\pi(f_{IF}/f_{DAC})n}\} = \mathrm{Re}\{j^{\pm n}x[n]\}$$

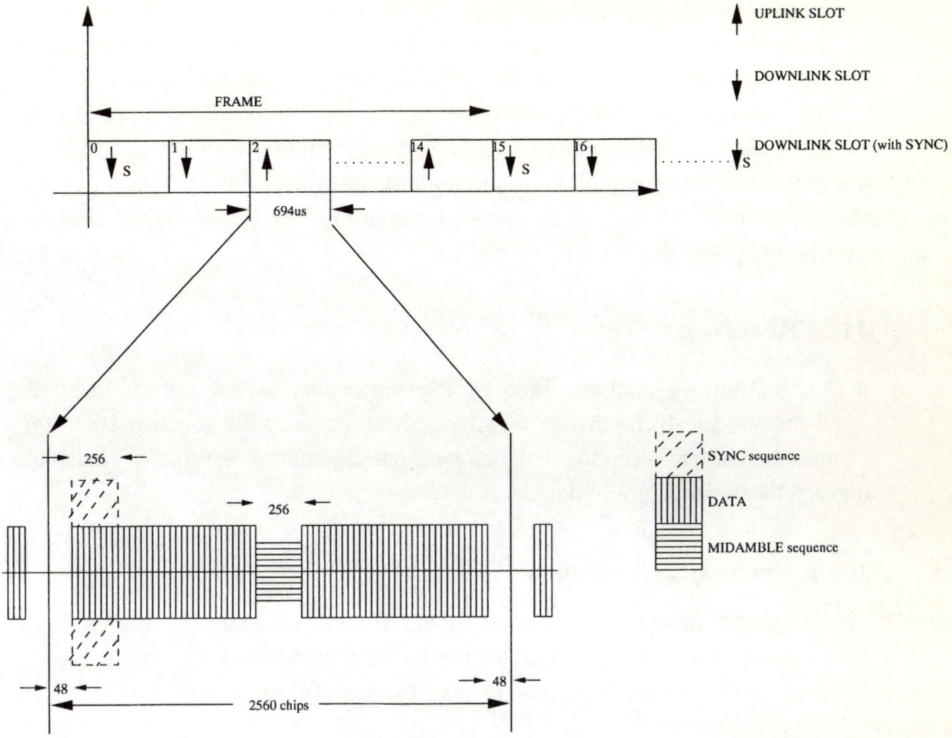

Figure 5. *Slot structure.*

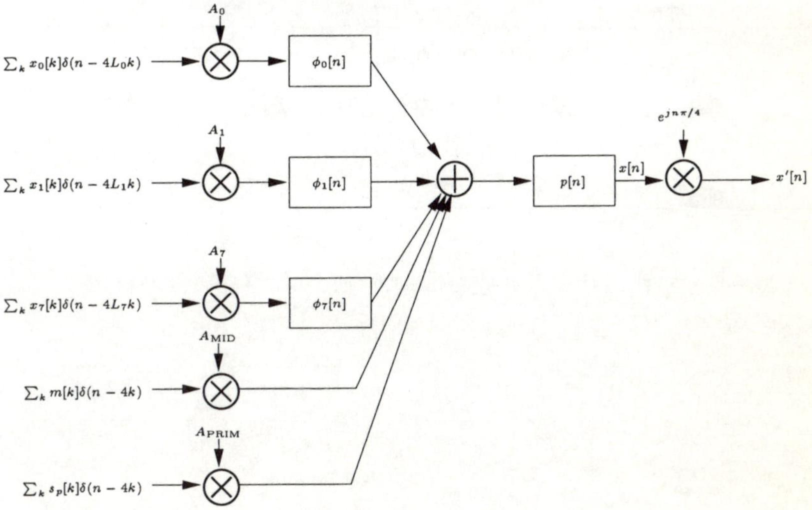

Figure 6. *Basic TX structure.*

In this way, the periodic spectrum of $x'[n]$ shows a spectrum replica centred at f_{IF} (see Figure 7(a)). After D/A conversion, a pass-band filter centred at f_{IF} removes the other replicas, generating the desired IF modulated signal.

The discrete-time modulation by $f_{DAC}/4$ in (5) requires a negligible computational cost since it corresponds to change alternatively the signs of $x[n]$ as can be noticed in expanding equation (5). In order to avoid aliasing when taking the real and imaginary part, the sampling rate must satisfy also the condition $f_{DAC} \geq 2W$.

III.3. Receiver

In this section we analyse some of the theoretical aspects of the receiver signal-processing. In particular we give a description of the receiver front-end architecture shown in Figure 10. Then we focus on channel estimation, matched filter synthesis and symbol detection.

III.3.1. Frame synchronization

Frame synchronization is achieved using a filter matched to the primary synchronization sequence to estimate the location of the start of a frame. This is achieved by filtering the bandpass received signal $r[n]$ as

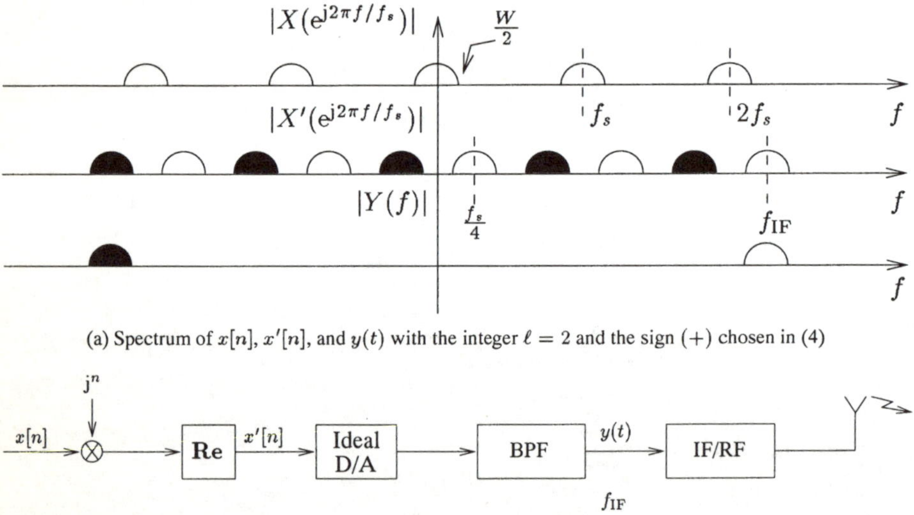

(a) Spectrum of $x[n]$, $x'[n]$, and $y(t)$ with the integer $\ell = 2$ and the sign $(+)$ chosen in (4)

(b) System model

Figure 7. Alternative approach for IF up-conversion.

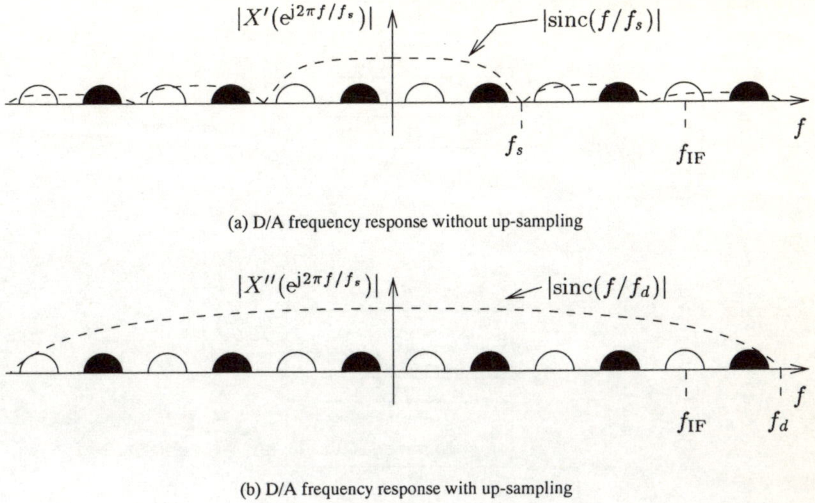

Figure 8. D/A frequency response.

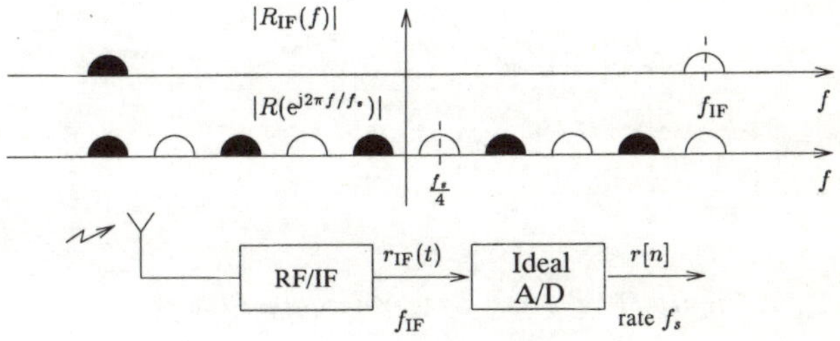

Figure 9. Receiver front-end.

$$(6)\quad r_s[n] = \left\{\sum_{k=0}^{255} s_p[k]\delta(n+4k)\right\} * r[n] = \sum_{k=0}^{255} s_p[k]r(n+4k)$$

and averaging the $r_s[n]^2$ over several frames. The maximum output of this filter is used to adjust the receive signal strength (via a variable gain IF amplifier) and synchronization is achieved when the maximum is greater than a pre-defined threshold. Note that this filtering operation involves purely real quantities. This is typically the most computationally intensive part of the receiver front-end since it requires a fairly long filter operating at the sampling rate. The 3GPP standard uses a hierarchical structure for the primary synchronization sequence which allows the filter to be implemented as a concatenation of 2 FIR filters of length 16. A block diagram of the primary synchronization sequence is shown in Figure 11.

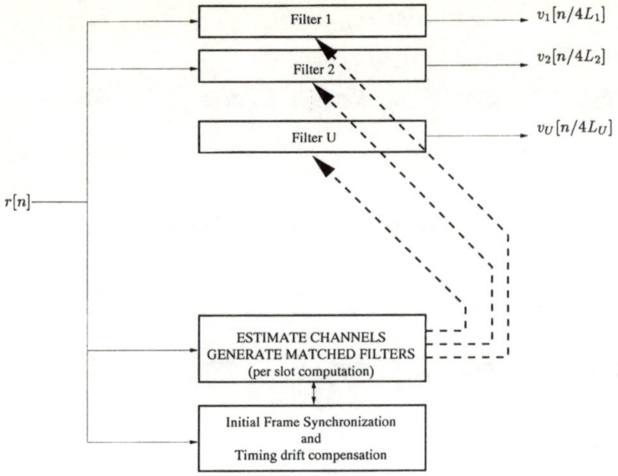

***Figure 10**. Basic RX structure.*

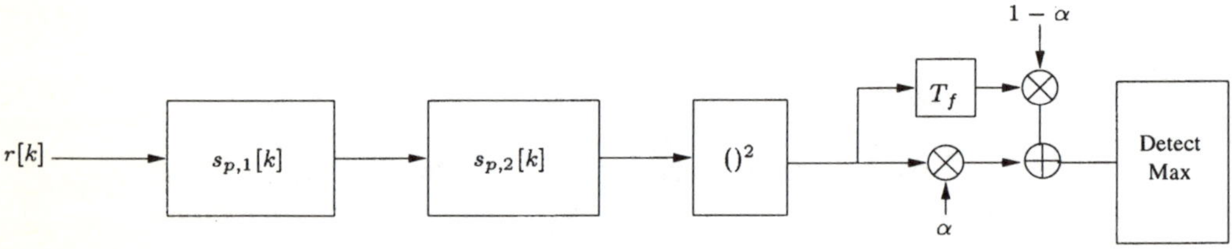

***Figure 11.** Primary Synchronization Processing.*

III.3.2. Channel estimation

Here we consider the training-sequence based multi-user channel estimation procedure for block-synchronous CDMA described in the UMTS/TDD standard.

In this scheme users are roughly synchronized to a common time-reference and transmit the training sequence at the same time (user timing errors are included as an effect of the channel and taken automatically into account by the estimation procedure). The maximum channel length (including possible timing errors) is Q symbols and the training sequence sent by each *u-th* user is a cyclic shift of the same common base training sequence $\mathrm{m} = [m[0], m[1], \ldots, m[M-1]]^T$ of length M symbols. This solution allows joint estimation of all user channels if $M \geq QU$, where U is the number of interfering users. It is proposed and described in [16, 17] and with some modifications in [18]. The interested reader is referred to these papers and references therein for more details.

Under these assumptions we can write the received signal sampled at frequency $f_{ADC} = N_c f_c$ during the M symbols spanned by the training sequence as

(7) $$w = \overline{\mathbf{M}}\boldsymbol{g} + \upsilon$$

where

(8) $$\boldsymbol{w} = [w[0], w[1], \ldots, w[MN_c - 1]]^T$$

is the received signal,

(9) $$\boldsymbol{g} = [\boldsymbol{g}_1^T, \ldots, \boldsymbol{g}_u^T, \ldots, \boldsymbol{g}_U^T]^T$$

is a vector containing the channel impulse responses of the U users,

(10) $$\boldsymbol{g}_u = [g_u[0], g_u[1], \ldots, g_u[QN_c - 1]]^T$$

is the u-th user channel filter vector and ν is a vector of interference plus noise samples, assumed to be white. The $MN_c \times MN_c$ matrix $\overline{\text{A}}$ is defined as

(11) $$\overline{\text{M}} = \text{M} \otimes \text{I}_{N_c}$$

where (⊗) denotes the Kronecker product and M is a circulant matrix containing all the possible cyclic shifts (by columns) of the base sequence **m**. The matrix $\overline{\text{M}}$ is also circulant and it is unitary similar [19] to the diagonal matrix diag($\overline{\alpha}$), where

$$\alpha = \underbrace{[\alpha^T, \ldots, \alpha^T]^T}_{N_c \text{ times}}$$

and where α is the Discrete Fourier Transform (DFT) of α. After some algebra [18], it is possible to show that the Least Squares estimation of the overall channel impulse response g is given by

(12) $$\hat{\boldsymbol{g}} = \text{IDFT}\left\{\frac{\text{DFT}\{w\}}{\overline{\alpha}}\right\}$$

where DFT and IDFT denote direct and inverse Discrete Fourier Transforms. The ratio of two vectors should be interpreted as the element-by-element division.

This approach can be applied to both base-band and pass-band signals. The receiver can also use the a priori information that the signal bandwidth is limited to W. Notice that this operation in the frequency domain corresponds to low-pass filtering in the time domain, moreover it reduces the computational cost since only a part of the MN_c products (by the element-wise inverses of $\overline{\alpha}$ in (12)) are computed. Eventually, after the IDFT, the processing gives the estimated channel complex envelope. The channel estimation procedure is summarized in Figure 12.

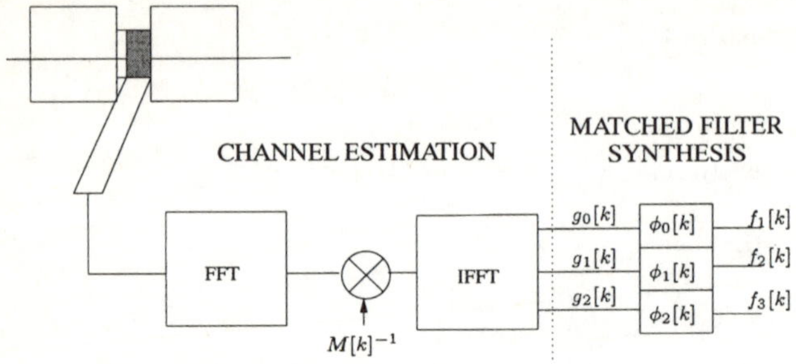

Figure 12. *Channel estimation and matched filter synthesis.*

III.3.3. Matched filter synthesis and data detection

Given the channel estimates, we are then interested in synthesizing a Matched Filter (MF) matched to the cascade formed by the user data spreading sequence, the chip pulse shape filter and the user channel. This is also shown in Figure 12. The overall impulse response for code i is given by

$$f_i(t) = \sum_{k=0}^{L_i-1} \phi_i[k]g(t - k/f_c) \tag{13}$$

where L_i is again the spreading gain. Using the sampled channel estimate $g[k]$ we synthesize the discretetime filter $f_i[k] = f_i(t)\,|_{t=k/f_{ADC}}$ matched to the overall response as follows

$$\hat{f}_i[k] = \phi_i[k] * \hat{g}[k] \tag{14}$$

In order to extract the data symbols we filter the received signal with the MF obtaining

$$v_i[k] = r[k] * \hat{f}_i^*[-k] \tag{15}$$

In this setting $r[k]$ *is* real while $\hat{f}_i[k]$ is complex so the product requires two real multiplications. To the contrary the baseband samples of $r[k]$ would be at half the sampling rate but would be complex. So the two complexities are identical.

Notice that the signal after matched filtering is still pass-band and the symbol estimates after sub-sampling are given by

$$\hat{b}_i[k] = v_i'[N N_c k] \tag{16}$$

where $v_i'[k]$ *is* the complex envelope of the MF output. But since

(17) $$v_i'[k] = (-j)^k v_i[k]$$

substituting into (16) and for $N_c = 4$ we get

(18) $$\hat{b}_i[k] = (-j)^{N N_c k} v_i[N N_c k] = v_i[4Nk]$$

In this way the symbol estimates are given by sub-sampling the MF output at symbol rate without taking care of the demodulation.

III.3.4. Carrier synchronization and decoding

The carrier synchronization is done at symbol rate with a classical decision directed algorithm [20]. The algorithm then takes a decision on the symbols and recovers the data (in our example a video stream).

III.3.5. Re-sampling

The baseband processing algorithms such as synchronization, channel estimation and data detection assume that the signal is sampled with an integer number, N_c, of samples per chip. In the current implementation we set $N_c = 4$. This solution avoids utilization of re-sampling techniques at both the transmitter and receiver front-end. These techniques have been studied in [15] and will be implemented in the next version of the software.

IV. Validation of the existing platform

The platform described in this paper has been validated by the transmission and the reception of two user full-duplex real-time video flows in an indoor environment. Two H263 video streams are transmitted in parallel and decoded in real time. For this we use the following parameters:

- spreading factor 16;
- bit rate 397 kbps (peak) per user;
- TDD configuration: 1 Tx slot followed by 1 Rx slot (transmission is done every 2 slots);
- two synchronous full-duplex streams per slot;
- RF band: 5 MHz at 2.1 GHz.

V. Conclusion

This first demonstration shows that the architecture of the platform is capable of sustaining real-time communications and is thus promising for future developments. The platform is currently being enhanced and opened to both industrial and academic collaboration. The enhancements will consist of:

- support for multiple antenna transceivers;
- more sophisticated signal processing algorithms;
- multi-user detection;
- layer 2 (RLC,MAC) functionality.

Collaboration has already begun under the label of the RNRT *(Réseau National de la Recherche en Télécommunications)* financing programme organized by the French Ministry of Industry and Finance. Three projects have been initiated covering the following topics:

- radio sub-system improvement (flexibility and sensitivity);
- compliance with the 3GPP UMTS/TDD specification;
- higher-level protocol stacks;
- integration to an IPv6 experimental backbone.

Acknowledgements

The authors would like to thank D. Margairaz and P. Favrat, formerly with STMicroelectronics, Geneva, for their collaboration in the development of the RF subsystem. In addition, the help of G. Vaucher and P. Bruchlmeier from ACORT/EPFL in the fabrication of many of the hardware components was greatly appreciated.

REFERENCES

[1] SRIKANTESWARA (S.), REED (J.H.), ATHANAS (P.), BOYLE (R.), "A soft radio architecture for reconfigurable platforms", *IEEE Communications Magazine,* (Feb. 2000).

[2] "Special issue on software radio", *IEEE JSAC,* Vol. 4, (April 1999).

[3] "Software radio", *IEEE Personnal Communications,* Vol. 4, (Aug. 1999).

[4] TSURUMI (H.), SUZUKI (Y.), "Broadband RF stage architecture for software-defined radio in handheld terminal applications", *IEEE Communications Magazine,* (Feb. 2000).

[5] ZANGI (K.), KOILPILLAIR (R.), Software radio issues in cellular base stations", *IEEE JSAC,* Vol. 17, pp. 561-573, (April 1999).

[6] RAPPAPORT (T.), Wireless Communications, *Englewood Cliffs,* (1996).

[7] 3GPP-TSG-RAN-WG1, "TS-25.2xx series", tech. rep., (Jan. 2000).

[8] DAHLMAN (E.), GUDMUNDSON (B.), NILSSON (M.), SKÖLD (J.), AB (E.R.S.), "UMTS/IMT-2000 based wideband CDMA", *IEEE Communications Magazine,* Vol. 36, pp. 70-80, (Sept. 1998).

[9] "http://www.itu.int/imt/2-rad-devt/index.html".

[10] FURUSKÄR (A.), MAZUR (S.), MÜLLER (F.), OLOFSSON (H.), SYSTEMS (E.R.), "EDGE: Enhanced data rates for GSM and TDMA/136 evolution" *IEEE Personnal Communications,* Vol. 6, pp. 56-66, (June 1999).

[11] MOGENSEN (P.), KREDERIKSEN (F.), DAM (H.), OLESEN (K.) , LARSEN (S.), "A Hardware Testbed for Evaluation of Adaptive Antennas in GSM/UMTS", Vol. *IEEE PIMRS'96,* pp. 540-544, (Oct. 1996).

[12] XILINX, "Xilinx 2001 FPGA Databook", http://www.xilinx.com, (2001).

[13] PLXTECHNOLOGIES, "PCI9080", http://www.plxtecj.com, (1999).

[14] 3GPP, "3rd Generation Partnership Project", http://www.3gpp.org, (2001).

[15] MONTALBANO (G.), NORDIO (A.), "Front-end software-radio review", tech. rep., Institut Eurécom, (July 1999).

[16] STEINER (B.), JUNG (P.), "Optimum and suboptimum channel estimation for the uplink of cdma mobile radio systems with joint detection", *European Transaction on Communications,* Vol. 5, pp. 39-49, (Jan.-Feb. 1994).

[17] 3GPP-TSG-RAN-WG1, "TS-25.221v3.1.0; physical channels and mapping of transport channels onto physical channels (TDD)", tech. rep., (Jan. 2000).

[18] CAIRE (G.), MITRA (U.), "Structure multiuser channel estimation for block-synchronous DS/CDMA", Submitted to *IEEE Transaction on Communications,* (July 1999).

[19] GOLUB (G.H.), LOAN (C.F.V.), Matrix Computation, *The John Hopkins University Press,* (1996).

[20] PROAKIS (J.G.), Digital Communications, *NY: McGraw Hill,* 2nd ed., (1989).

[21] "Spectrum Signal Processing", http://www.spectrumsignal.com, (2001).

[22] FSMLABS, "RTLinux", http://www.rtlinux.org, (2001).

Chapter 8

Wireless communications+++

R. Steele

Multiple Access Communications Ltd, Southampton, UK

I. Setting the scene

Some five years ago the writer asked groups of experts in mobile radio communication what would they do after they achieved their cardinal goal of *communications at any time, from anywhere, at any speed.* It transpired that, in general, the question was unwelcome, made them feel uncomfortable, was considered irrelevant, and for some there was nothing after the great goal had been realised. That was only five years ago! Now many people have thought and published treatises on the next phase, and it appears that the notion of multimedia services, particularly Internet Protocol (IP) services, while a user is nomadic or mobile is what the industry has girded its loins to achieve. So before we have achieved our first great goal we have modified it, changing our view of our network from a vehicle to support mainly voice communications, to one that provides communications and information when and as required with aspirations of cyberspace. We will be physically in one place at an instant in time, but electronically we may access remote libraries, participate in virtual businesses, experience thrills without getting out of our armchairs, and so on. And then what? To where are we rushing, and will we be happier when we arrive?

This short discourse is concerned with where mobile communications is heading, and we will return to some of the philosophical issues raised above at a later stage. Our approach is to start by considering communications for the first decade of the 21st century. This is not easy as unforeseen profound events are likely to occur, but we know that third generation (3G) systems will be deployed, and that during this decade they will evolve. After that we will attempt to move to later decades. Prediction so far in the future formulated on what is happening now will be very unrealistic, and instead we will apply preconception based on how society might be evolving to guess what future networks might be like. Acknowledging at the outset that we will fail in this objective, we may succeed in initiating a serious debate on the long-term consequences of our actions now and in the immediate future.

II. The first decade

In the 1970s and until the first half of the 1990s mobile radio for an engineer new to the subject was dominated by understanding radio propagation and witnessing furious debates as to the suitability of different multiple access methods. The radio interface between users and the network and what was required by the mobile component of the fixed network became clear. The fixed network needed special sub-systems, such as databases, switches, gateways, and billing systems, and inter-working units with the networks of the main telecommunication companies. Originally most calls were between a mobile user and a PSTN/ISDN user. The radio interface had a specific multiple access arrangement and a modified open systems interconnection (OSI) reference model that was predominantly concerned with the lower three layers of this model. Most calls on first and second generation networks are voice calls, with some low speed data offering short message services (SMS). Recently there has been the introduction of the wireless application protocol (WAP) to provide web-like services, and the very successful iMODE services in Japan.

By the commencement of the first decade of the 21st century the de facto 2G world standard, GSM, is beginning to have its Phase 2+ enhancements [1]. There will be the general packet radio service (GPRS), which will support packet data rates in excess of 150 kb/sec; high speed circuit-switched data (HSCSD), which uses TDMA slot aggregation to increase the circuit-switched data rate to 57.6 kb/sec; and the enhanced data rates for GSM evolution (EDGE), which applies multi-level modulation to GPRS and HSCSD. In addition there is the cordless telephony system that provides a GSM home base station; a subscriber identity module (SIM) tool kit that allows standardized access to SIM-based applications; a standardized intelligent network environment, known as CAMEL; work-group services including broadcasting and group calls and so on. The consequence is that the GSM backbone network is required to handle both packet-switched and circuit-switched data, to operate at much higher speeds and capacity, and to offer a wide range of services. From a service point of view GSM is becoming a 3G network.

While this Phase 2+ programme is accelerating, the Europeans decided that their 3G system, called the universal mobile telecommunications system (UMTS) would have a radio interface that would be either wideband CDMA for use with paired spectrum, or a TDMA system with multiple user CDMA signals in a slot for unpaired spectrum. The wideband CDMA system therefore uses frequency division duplexing (FDD), while the TDMA/CDMA system employs time division duplexing (TDD). These UMTS interfaces are radically different from those of GSM Phase 2+. However, the radio interfaces of both GSM Phase 2+ and UMTS will plug into the same enhanced GSM backbone core network, whose architecture is displayed in Figure 1. We observe that the conventional GSM network subsystems, such as the BSS, MSC, VLR, HLR, etc, are present, plus elements to support GPRS. The UMTS terminology for a BSS is a radio network sub-system (RNS), and for a BS it is a Node B. Figure 1 shows a RNS with interface I_{uCS} for circuit-switched data from the radio network controller (RNC) (the UMTS version of a BSC) to the MSC; interface I_{uPS} from the RNC to the packet node SGSN; interface I_{ur} between RNCs; and interface I_{ubis} between the RNC and the Node B. It may be expected that with the advent of software radio the BSS and RNC will be a single entity, as will the BTS and the Node B. We may therefore conclude that our networks will evolve to handle both packet and circuit-switched data and with fibre/radio nodes which can adapt to GSM, and UMTS as well as a range of other radio interfaces. Packet transmissions may eventually be exclusively preferred over circuit-switched transmissions.

This approach of a core network handling different radio interfaces is not new. IS-95, now cdmaOne, another global standard, was conceived as a radio interface, and its original version was a twin of AMPS in a dual-mode mobile. As IS-95 was CDMA and AMPS analogue FM using FDMA, the new digital system had to work on the IS-41 AMPS backbone network. Nowadays cdmaOne mobiles are often single mode, operating in both the 800 and 1900 MHz bands. The evolution of cdmaOne to the 3G system cdma2000 is relatively straightforward. cdma2000 will be either wideband CDMA spanning a bandwidth of multiples of 1.25 MHz, or it will use multi-carriers, where each carrier has the cdmaOne 1.25 MHz channel occupancy. Thus the lowest state of cdma2000 is cdmaOne (but an enhanced version).

The expectation of a single global 3G standard has not materialized. Instead the ITU calls its IMT-2000 a family of 3G systems that will be inter-worked [2, 3]. The terrestrial component has five systems; the two UMTS systems called IMT DS (direct sequence) and IMT TC (time code); the cdma2000 system called IMT MC (multi-carrier); IMT SC (single carrier) which is better known as EDGE; and strangely IMT FT (frequency time) which is the DECT system. There is also the mobile satellite system (MSS) component to provide global coverage at low capacity.

Harmonization is in progress to inter-work UMTS and cdma2000, and as EDGE is on the UMTS road map, then early software radios (which will not include down-conversion using a DSP) will be able to combine these systems. Further, MSSs, like Globalstar, are able to inter-work with both GSM and cdmaOne. So the realization of a single radio interface that can adapt to most cellular systems is expected during the first decade.

An interesting question relates to operators who have, say, GSM spectrum in one band for which they pay a modest licence, and 3G spectrum they acquired at huge cost in a spectrum auction. How will they use these bands when the mobiles can tune to either band, or perhaps to both bands at the same time, using software radio? Will they have speech and low speed data on the old 2G bands, only high bit rate services on the 3G bands, and will they be allowed to acquire small pieces of spectrum in other bands where they could offer private mobile radio/special mobile radio (PMR/SMR) services? Will they operate in unlicensed bands? Will regulators grant them the freedom that they might expect of the market place? Attempting to accommodate in cellular structures roaming users having bit rates from a few kilobits per second to 2 Mb/sec with different quality-of-service (QoS) agreements in one spectrum band is a task of monumental difficulty. Network design is significantly simplified if similar bit rate/QoS services are grouped together in different bands. There is another key question that should be asked on this issue, and that is, should operators be forced to share bands? By band sharing a wider bandwidth is available and hence higher bit rate services can be offered. If bands are shared, other than

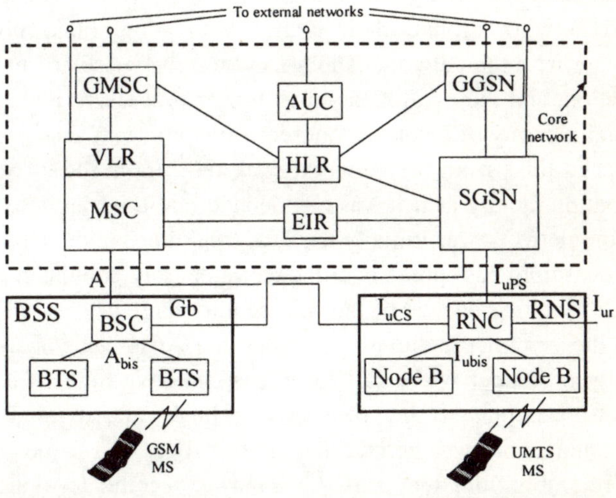

***Figure 1**. Network architecture for GSM and UMTS and their evolution.*

AUC = Authentication centre
BSC = Base station and controller
BSS = Base station system
BTS = Base transceiver station
EIR = Equipment identity register
Gb = Interface between BSS and SGSN
GGSN = Gateway PPRS support node
GMSC = Gateway MSC
HLR = Home location register
MSC = Mobile switching centre
RNC = Radio network controller
RNS = Radio network sub-system
SGSN = Serving GPRS support node
VLR = Visitor location register.

introducing fairness algorithms to make sure an operator is not gaining an unfair advantage, we need to develop very powerful anti-interference/interference avoidance algorithms. It is very difficult to predict in this area because of the current serious consequences affecting the industry caused be auctioning spectra.

Most commentators and industrialists are betting that IP services to mobiles is the big growth industry for the first decade. Since they all seem so confident of their predictions they must have done a lot of market research, or if you are a sceptic, a lot of navel gazing. During the first decade the bandwidth will limit IP services, unless a user goes to nodal cells [4], i.e., high bit rate "watering holes". To achieve widespread use of 2 Mb/sec services, huge amounts of infrastructure will be needed so that nomadic and stationary users will always be close to a picocell. For others, IP services will probably operate at 64 kb/sec or 128 kb/sec (it may be difficult to accommodate roaming 384 kb/sec users), and for such services to succeed they must be *personalized*, e.g., personal navigation, observation of your house for security reasons, monitoring your children at play, etc. Video is therefore a vital service, but it will be of relatively low quality, except for those at the "watering holes".

With the advent of body wireless area networks (B-LANS) users will have a distributed network around their bodies with some radio nodes that will communicate with the network. The B-LANS are the successor to the hand-held mobile station, and they will have much higher processing power and memory storage, and an array of displays. Consequently, non-real time IP services could use low speed links from the network, store the data in the B-LAN, and display the information rapidly as if it was a wideband real-time service.

The fibre network was initially used for trunk lines in the fixed network, but slowly optical fibre has spread over the entire densely populated regions. At the beginning of the first decade of the 21st century there are few fibre lines to the home, but the asymmetric digital subscriber line (ADSL) is being used to increase the capacity of copper wires so that homes can have higher bit rate services. However, fibre is innately inexpensive with huge capacity, and as the decade progresses more fibre will be laid directly to offices and homes. Whether it be ADSL, or preferably fibre, that is used, we may expect the developed countries to *wire-up* their urban areas, and in later decades their rural environments. This is not an over-ambitious task as countries have already been wired for electricity. As this telecommunications wire-up gathers pace, what is the role for mobile radio communications? With fixed high capacity links in every building we will be able to form very small radio cells. The base stations will often be of less than a matchbox in size, radiating very low power. All of us will live near a base station and none of us will care! No-one will say they are dangerous, nor unsightly as they will not be visible, being embedded in building and city infrastructures. With base stations (BSs) costing very little, they will be deployed ubiquitously along with cameras, and whether they will be used regularly or not will not be important. They will be installed for functionality and convenience, in the same way that lighting and power outlets are used.

Let us pause to consider the implication of myriads of small BSs and their accompanying MSs radiating low levels of electromagnetic energy over a wide frequency range and very small coverage areas. The result will be that people will be exposed to frequent low level electromagnetic fields. This is in contrast with the current situation of few BSs and MSs transmitting at much higher power. Given this the expectation is that the average exposure in the future will be less.

Along with the miniaturization and widespread deployment of BSs will go the installation of physically small switches/routers and databases, although this may not happen until the second decade. Gone will be the concept of accessing remote massive switches and databases. Decision making regarding the efficient use of resources, particularly radio resources, will need to be fast and tailored to local events. Knowledge of available resources, where the mobiles are currently located and where they are predicted to be going, QoS agreements relating to the active mobiles, current link qualities of all links, and so on, will be locally utilized in deciding what actions to take. Local mobile networks will have generic similarities with campus networks having private branch exchanges, although the switches and databases in future mobile networks will be small and more localized, e.g., a switch/router in an office block or large office. So with microcells go micro-networks with their small components. Compared to the present day the equipment effectively disappears in that it will be coffee mug size.

We mentioned that many BSs will be co-sited with cameras. Tiny in size, the cameras will be difficult for people to spot their presence, but they will send and store images of us. While the ubiquitous camera will be an important aid in decreasing crime, and allow us to observe events at any distance, they could also be used by governments to ensure that we obey their wishes.

The small radio cells will be used in numerous ways. There will be wireless local area networks (W-LANs) that will connect home appliances and entertainment systems and interact with a user's B-LAN. Entertainment will follow a user around, appliances will be remotely serviced, and complex security systems will be a growing business. Whenever possible, people will use the fibre network for accessing the Internet, avoiding the mobile component with its limited bit rate. Mobile Internet will be for personalized services when nomadic or on the move. Logging onto the Internet will need to be much faster and browsing more effective. Users will prefer closed community networks enabling them to use services, do shopping, etc, with the feeling of security, and with a quality-of-service they find acceptable. These closed networks will enable employees to work at home, or elsewhere, creating secure virtual businesses.

III. After the first decade

Predicting events is very difficult, predicting when events will happen is even more so. Accordingly, we will discuss possible events without saying when they might occur. Let us start by asking what will happen to our existing technologies.

III.1. Winners and losers

There are numerous radio systems that have appeared at a time when there was a social need for them. The social need is like a time window over each technology, and eventually the technology slips out of the window and is no more. The first mobile radio system was wireless telegraphy and was much needed by society. In particular it enabled communication with ships as they moved around the world. This technology relied on ground stations. It has now gone, superseded by single sideband (SSB) communication via the ionosphere, which in turn was replaced by satellite communications.

International voice communications were handled by submarine cables (point-to-point) and then wireless-voice communications via the ionosphere enabled any point-to-any point communications. Nowadays the fixed network spreads across the globe via a mesh of connections, and for inaccessible locations communications are provided by MSSs.

Television broadcasting had a dramatic effect on peoples' lives, changing social habits, opinions and behaviour. The transmissions were placed in the VHF and UHF bands that, at that time, were sparsely occupied. Now these bands are coveted by mobile operators because they support their cellular communications. As a consequence, the broadcasting companies will eventually be forced to deliver their services via fibre and satellite. Broadcast companies will not be willing to part with *their* spectrum (and they do think that nature gave it to them!) and will try to provide value added services by using broadcasting for high bit rate downlink transmissions, and low bit rate uplink transmissions using other networks. However, those services that utilize broadcasting channels can never be as well targeted at individuals as those that can be provided via mobile networks or by fibre networks, so television broadcasting in the VHF and UHF bands will be one of our losers. Broadcast services such as sport, news, weather, will continue, delivered by GEO satellites or fibre networks, but they will not be in the mobile frequency bands.

Private mobile radio (PMR) has been used by special groups, such as the emergency services, police, and transport companies, for many years. PMR was the precursor to cellular radio. It lingers on into the age of cellular with insufficient spectrum and without the huge market size of cellular that forces down the costs of mobile stations and network infrastructure. As cellular becomes more ubiquitous, inexpensive and with a wide range of services, it will create virtual PMR networks within its cellular architecture. The result will be that conventional PMR will cease.

Wireless local loop (WLL) delivering point-to-point services to the customer via wireless from a BS, has a continuing role to play in developing countries where installing fibre infrastructure can be difficult and/or expensive. In the developed world, WLL will linger in the USA, but eventually it will be a blip in history. It will die because of spreading fibre infrastructure and the wide use of cellular radio.

While satellite systems have a long term role in monitoring and surveillance, they have only a medium term existence for communications (as distinct from TV broadcasting). They will be displaced by aerial platform technology and other space systems as we will discuss later.

Of the systems existing at the beginning of the 21st century the one that will last the longest is cellular. It is the king of the wireless world. Yet as time goes on most cellular communications will not be between people, but between machines.

III.2. Releasing radio spectrum

We will now ponder how much bandwidth we might ideally need for mobile services. The answer is, never enough. This is the opposite situation facing optical fibre engineers who ask what will we do with all the capacity we have.The answer is, wait and we will tell you. But, mobile users must be given more bandwidth if high bit rate services are to be supported. So what bandwidth is potentially available? The criterion might be that the transmission frequencies must be able to significantly propagate around corners.

We note, however, that for high frequencies where diffraction around buildings is essentially non-existent, then cellular systems can be implemented [5], but only by the deployment of massive numbers of network nodes. Imagine a building where every confined space has so many small fixed nodes, i.e. BSs, transceiving at low power and at frequencies of, say, 60 GHz, that total coverage is achieved for any mobile node, ie MS. As office and home environments will have many more fixed nodes than mobile nodes then the utilization of line-of-sight frequencies becomes an attractive option, as the available bandwidths are comparatively huge. This technique, of using line-of-sight transmissions in a cellular mode, will eventually be spread to many outside locations, e.g., in street canyons.

Returning to the frequencies that have good diffraction and refraction properties around and within buildings, suppose that 5 GHz of bandwidth (virtually the first 5 GHz of the radio spectrum) is available. Obtaining this bandwidth is a political job for the ITU's World Radiocommunication Conference (WRC). Suppose in any one location there are five operators, and that each has 1 GHz of bandwidth. Each operator would need to decide how many carriers it would deploy, and the type of multiple access, modulation, etc, but carrier spacing of 100 MHz or more would be possible. Now mobile users would have tens of Msymbols/sec available which might satisfy the most of their needs. How an operator would acquire such a large bandwidth is a matter of conjecture. It should not be by auction, which is no more than a government tax on mobile users, as the amounts of money would be too great. Perhaps we will return to the sanity of the beauty-contest in allocating radio spectrum.

An alternative way to use a large increase in the amount of spectrum is to divide it into bands where each band offers a different bit rate service [6]. There will be a proliferation in the number of service providers (SPs) who will offer their services on numerous networks. If a SP offers interactive videophone services, say, that require high bit rates, low delay, and low bit error rate (BER), then the SP will negotiate with each network operator to ensure that sufficient infrastructure and functionality exists in their network to satisfy the QoS the SP will offer to its clients. The SP could also offer subscribers different charges depending on the QoS, which might mean that different bands are used for the same service but the grade-of-service (GoS) may be different as one may be far more congested than the other band. There will also be a *yellow pages* web site where a subscriber can go (while mobile) to find a SP that provides the type of service required. Further, a subscriber will not be limited to one network operator. A MS on switch-on will scan all the networks, camp on one, find a suitable SP and get a service. Incoming calls will go to an organization (could be a network operator) with a HLR to which the user subscribes. Via this HLR, communication with the VLR of the network on which the user has camped on will ensue, enabling the user to be paged from this network. Billing will be via specialist billing agencies.

III.3. Going soft

As software radio interfaces (SRIs) [7] become able to cover a wide frequency band(s), to reconfigure their transceivers over a packet duration, to house the necessary IN, and have personalized software; then so will the backbone network become soft in that it will be a reconfigurable ad hoc network that will respond locally to local needs, while being able to establish intercontinental connections. A user with his B-LAN will have a SRI to communicate to the network fibre/wireless hybrid node (a BS). The B-LAN will be a complete network distributed about the body, having fibre in the clothing linked to a power source(s) charged from body energy. This B-LAN will have considerable processing power and memory. B-LANs will communicate with each other, to form ad hoc wireless networks, as well as with the numerous network nodes of home, office, street, MSSs, etc networks. Figure 2 shows a fibre network, nodes and radio cells.

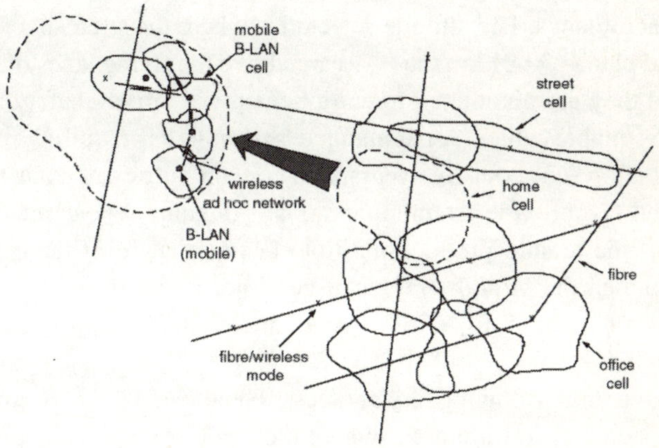

Figure 2. *Fibre network, different radio cells and an ad hoc B-LAN network.*

III.4. Air supremacy

The MSS, ICO, Globalstar, and the now defunct (?) Iridium system, were IMT-2000 proposals. They are low earth orbit (LEO) systems, and therefore enable relatively small hand-held terminals to be used. They are also multi-mode terminals and can be used on terrestrial networks. 3G terminals will eventually be able to communicate via all members of the IMT-2000 family of MSS and terrestrial networks. We may also anticipate that 3G terminals with their soft radio interfaces will be able to handle signals from broadband satellite networks, such as Teledesic and Skybridge, as well as from other satellite systems in higher orbits, although enlarged terminals will be required for non-LEO MSSs.

MSSs do provide global coverage (although the loss of Iridium will make communications with mobiles in the polar regions difficult), but they also create large cells on the earth's surface with their multi-beam antennas because they travel at a height of several thousand kilometres. The high altitude of flight also results in a substantial delay, and the speed of LEO satellites causes large Doppler shifts. Their link budgets are such that line-of-sight radio links are preferred, and communications from within buildings is not, in general, supported. Furthermore, satellites are expensive, their launch costs are high, and they cannot be retrieved from space for servicing and technological updating. All these disadvantages are tolerated because of global coverage.

MSSs will therefore continue to be used until there is a better alternative to provide coverage and capacity in areas where it is too expensive to deploy terrestrial cellular systems. Fortunately there are alternative methods. One that is actively pursued today are high altitude platforms (HAPs). "High" in this context means the stratosphere, typically 21 km, and the platforms are either airships or aeroplanes. The altitude selected is where the stratospheric winds are minimal and planar. At 21 km the HAPs are above the flying lanes of commercial aircraft, and they are untethered to earth because of aircraft safety and because there are no suitable cables. An unmanned airship HAP is required by proponents to stay on-station for about five years, and convert sufficient solar energy to do this, as well as provide communications. Antenna arrays attached to the underside of the airship produce multiple beams that form radio cells on the surface of the earth. Aircraft HAPs can be either manned or unmanned, fly in tight circles of some 5 to 8 miles in diameter, and carry arrays that form terrestrial cells.

HAPs have the following advantages. They can be sited over cities to increase capacity, or in rural areas where they can provide wide area coverage. Each HAP can produce tens to hundreds of cells and these cells can be tens of kilometres to a few hundred metres [8–10]. HAPs are linked to ground stations and to other HAPs to form HAP sky networks. The cells can be instantly moved by electronically changing the direction of a beam. Compared to MSSs, the signal delay of HAPs is small, they do not contribute to the Doppler frequency, the radio link budget is much better, the spectral efficiency is enormously higher as the cells are small; they can be returned to earth for technology updates; they are relatively inexpensive to make and launch, one HAP can produce a cellular network over hundreds of kilometres, and so on.

The aircraft HAP is the HAP likely to be deployed initially, and as a consequence, the antenna structures cannot be too large. This means they will operate at relatively high frequencies (e.g., providing local multi-point distribution service (LMDS)) or if in the 3G bands, they will only create relatively large cells. The current strategy is for aircraft HAPs to be piloted, and for crews to do shifts on-station. This means that for one location, three crews and two to four aircraft are required. Aerial refuelling of aircraft HAPs facilitating long flights are not proposed. Flight regulations are not an impediment for aircraft HAPs.

Airship HAPs offer a better long-term solution (at lower latitudes) as they are some 200 m in length, can have huge antenna arrays and therefore can create microcells at 3G frequencies. They must, however, have stabilizing systems, particularly to compensate for tilting of the antenna arrays, and be held on-station as an operator cannot have a network gently floating away!

Other aerial platforms (APs) will be used. Some will be housed on extraordinarily high structures, where they will create multiple cells that move around as required. Others will be on tall towers and tethered balloons. Many civil aviation aircraft could have APs to form terrestrial cells as they go about their business of delivering passengers to remote destinations. Some APs will be small and attached to unmanned aircraft that fly to teletraffic hot-spots, etc. APS can be at any height, subject to aeronautical constraints. The APs will mark the demise of MSSs. We might also speculate on the longevity of APs, a technology that has not yet arrived. With the future ubiquity of fibre networks augmented by radio tails, the effectiveness of AP technology will decline, but it will always have the feature of providing many cells at switch-on whose characteristics can be controlled electronically from the ground.

However, communications from space will continue, based on the large orbiting space stations that were beginning to be assembled at the end of the 20th century. We may envisage a network of these space stations, descendents of MSSs, orbiting the world, of huge size, monitoring earth as well as space, and as a sideline, providing communications [11]. The problems of Doppler shift and delay associated with MSS would continue, but the spectral efficiency would be high as the space stations would form very small cells using very large antennas. This space station system (SSS), with its orbiting space stations interconnected via laser beams or high frequency radio links, would be able to deliver mobile services, very high bit rate point-to-multipoint communications, and broadcast services. They would also interwork with the aerial platform layers and terrestrial network beneath them.

III.5. From beneath the waves

Numerous optical fibre communication links criss-cross our oceans carrying vast amounts of traffic. Some of this traffic can be used to communicate with ships sailing on the oceans and with transatlantic aircraft. To achieve this, sea platforms can be housed on unmanned surface vessels, powered from the waves, and connected to the submarine cable beneath the waves by another fibre cable. By this means a sea platform taps into the submarine cable. A particular submarine cable will have cables to sea platforms at numerous points along its length. Antennas connected to a sea platform produce multiple beams, i.e., cells, for ships in their locality and over-flying aircraft. As the submarine fibres are laid in many directions the surface platforms connected to them are able to provide continuous coverage across an ocean. By this arrangement we have maritime radio cells for ships and aircraft communications connected by a high speed, high capacity backbone fibre network.

IV. Where is the network heading?

The invention of language was a major step in man's evolution, and the ability to hold a conversation with a person thousands of miles away was a monumental achievement. Our first global network was a point-to-point one, then came the provision of mobility, and later low speed data services. The network has evolved from an analogue one to a digital one where all signals are transmitted as a humble bit. Some bits are more important than others, they do not travel at the same speed, nor do they experience the same delays, nor travel via the same media. The bit is the atom of communications systems. It can have one of two voltage levels, and these and its duration are system parameters. It can also be viewed abstractly as a logical entity; a logical 1 (true) or a logical 0 (false), and using Boolean algebra we can construct processors, which themselves process bits. The bit is both carrying information, and in its logical representation, being manipulated to carry out functions. Scaling this process results in the code conversion of complex entities, such as the signal from a camera, to a bit stream, and its distribution to where it is required. It enables us to not only have communication networks, but to use them to gather and distribute information, entertainment, and to produce cyberspace. The bit would appear to be here to stay.

The applications required by our network will become myriad. Our homes will be micro-networks with wired and wireless LANs monitoring and controlling how we live, from checking the status of the milk in the refrigerator, to servicing the washing machine, to presenting pictures of visitors at the front door. We require to see our house and its contents any time from anywhere. To achieve this people will use their B-LAN. There will not be a problem with battery power as our bodies generate some 100 watts and we will harness a fraction of this power for our body network. This network will have huge processing power and memory along with holographic and other displays, voice inputs, etc, using biodegradable electronics. There will be other similar networks in offices, on the streets, and even in remote areas. Ad hoc networks that are adaptive and responsive to local demands will be common, but there will be ad hoc network structures in a hierarchical order having lower order ad hoc networks embedded within them.

All of these networks, from the international to the one in the refrigerator, will be soft, i.e., adaptable and flexible to requirements and demands. Such flexibility, particularly to unforeseen events, requires complex software, software that is itself changeable. In essence software that, when in a crisis, will adapt itself. This means code that writes code (we already have this to a limited extent). These codes will be the software agents (SAs) of the 21st century, the atom of the software world, the oil that lubricates our interconnecting software engines.

Let us examine the role of SAs. As an example, consider a B-LAN. When the wearer speaks, the human interface SA (HI-SA) will, say, visually appear on one of the user's display systems and ask what duty is requested. Suppose the wearer is interested in acquiring an antique dressing table. The HI-SA will gather details for the purchase: the origin of the table, its period, cost, size, when it is needed, and so on. The HI-SA will have a mobility SA (M-SA) make an access by radio to a nearby network, visit the yellow pages web site, and obtain a list of service providers (SPs) who deal in antique dressing tables. The M-SA will then visit the SAs of the SPs, eventually do a deal, and report the outcome to the user. Suppose on another occasion the wearer asks the HI-SA what precautions he should take in climbing a ladder to examine his chimney pot. The HI-SA will ask for information, and then may need to have a SA devise and use basic equations involving the theory of statics. If a program is unavailable, the SA will have to have it written, or ask the M-SA to get either the equations or the result via a SP. Generally the B-LAN will have the code written locally by body SAs, only approaching a SP when solving complex problems.

Currently there is much discussion regarding software down-loading to terminals. Some engineers consider that the terminal should be devoid of much memory, and consequently the same information will be frequently down-loaded when required. However, future B-LANs will have formidable storage capacity, and once acquired, software will be stored locally. This procedure is also in accordance with people's acquisitive behaviour, e.g. many people prefer to buy and keep their own books in preference to using a public library.

The network will use SAs to make decisions, and to control, maintain and adapt their networks. At one extreme we have experienced computer viruses, which we may think of as malevolent SAs. The SAs created for our networks will be created by benevolent *software robots*. They will not, however, be the equivalent of robots on a car assembly line, programmed to undertake specific tasks, although they may start off that way. SAs will be large, mobile, autonomous code and will be created by different engineers with their personal preferences. They will be conceived to undertake tasks, and inevitably they will have human-like characteristics. For example, in buying an antique dressing table the SA will need to be told a lot about the subject as well as handling business matters. In other words we will need to impart to SAs a lot of human knowledge, expectations and acceptable goals. Since we also require them to act independently to unforeseen events that must be dealt with immediately, they will probably develop genetically. The SAs will therefore occupy a software world that is psuedo human, and hence SAs will not be good or bad, but have complex behavioural patterns that will sometimes do things to the networks that we do not expect.

Although human intelligence evolves extremely slowly, we have been spectacularly successful in using it to construct modern civilization, and now we are beginning to understand ourselves. However, we still know relatively little about our brains and minds. When we consider our network, its processing power, memory, reconfigurability and evolution are increasing rapidly. When we apply the word *intelligence* to the network, or machines in general, it illicits a howl of protest. This appears to be because intelligence seems to be deemed an exclusive human attribute, and according to this way of thinking, if a machine does not have what we think of as human intelligence, it is not intelligent.

You could argue that man has tried to imitate birds, in learning to fly, but has produced aircraft that cannot flap their wings, dive underwater to catch a fish, etc. Of course, we did not want to do these things, and we can point out that birds cannot fly faster than the speed of sound nor carry hundreds of other birds over thousands of miles. Although man and birds both fly, they do it differently to suit their respective needs. Similarly, machines and man will both be intelligent, but in different ways according to their makeup and the goals they need and want to achieve. So with the passage of time our networks will have highways carrying enormous quantities of traffic at huge speeds. They will have SAs to allow them to be self-organizing and reconfigurable. Decisions and executing tasks will be done at a formidable rate; vast numbers of interconnecting machines will be accommodated; there will be rising complexity in service provision, and so on. The SAs will create a software world that will be intelligent.

If this is a glimpse of the future, it follows that we may at this time be launching a *global brain* of sorts. This is not like the brain of an individual. It is more akin to a complex society of interconnecting happenings on many levels; a representation of our world but in the software domain, the so-called cyberspace. The hardware of our network, with its multiplexity of wireless and optical nodes, will somewhat resemble the network's *brain*, while the SAs will give it the *societal mind*.

Industrial strikes by key workers, or picketing of key installations, such as oil refineries, can bring our societies to a halt in a matter of days. People are unable to drive their cars to work or to the shops, staff and patients cannot get to their hospitals, food distribution ceases, and so on. Laws that were fashioned for the 19th and for most of the 20th century seem inadequate to handle the types of crises that can now suddenly erupt. These *flash crises* that can bring a country to a standstill in a few days will be trivial and slow to occur compared to what may happen in the future when society will be dependent on trillions of interconnecting processors, SAs, complex ad hoc networks, applications running at high speed, and so on. If some SAs are forced to make rapid decisions in response to some unforeseen events, and their decisions result in an almost instantaneous breakdown of the network, the equivalent of a brain seizure ensues. Society that will now be so dependent on the brain for almost every type of decision making, production, distribution and control, will be in a catastrophic state. In the same way that bulkheads compartmentalize ships so that in the event of an accident only some parts of the ship will be flooded, we need to design our networks with software bulkheads. This task should be a major research activity. A serious fault in the network needs to be rapidly isolated, rather like we isolate a person with an infectious disease, and dealt with. The "brain" that will make life for most citizens easy and enjoyable, must be self-healing.

But, will we be able to design and control our global brain with the passage of time and its growing complexity?

Mankind is faced with a dilemma. As our science is becoming increasingly complex as time goes by we have fewer people able to understand it. For example, what proportion of people will be able to comprehend the leading edge of computer science in fifty years? Very few. Hence we may be inclined either to do little and arrest mankind's *full-ahead* mentality, or we go full-ahead knowing that we risk losing control of our increasingly complex civilization. There is, however, a way out. Full-ahead to the global brain, but being able to control it because we biologically change ourselves. We institute changes to rapidly evolve our own intelligence to keep ahead of our machines. This involves knowing ourselves in a detailed manner, making biological changes, and introducing internal body processors so that we can make rapid and complex decisions. *Homo sapiens* then forms the genius society and the machines remain our servants.

Oops, we have gone from 3G to *science fiction*, or have we?

Acknowledgment

The author thanks the organizers of the ECWT 2000 Conference in Paris, 5-6 October 2000, for permission to amend the paper he presented there, entitled *Full ahead to where?*, and to publish it here.

REFERENCES

[1] ZVONAR (Z), JUNG (P.), KAMMERLANDER (K.), "GSM – Evolution Towards Third Generation Systems", *Kluwer Academic Publishers*, Boston, (1999).

[2] *IEE Conference on 3G Mobile Communication Technologies*, London, (27-29 March 2001).

[3] STEELE (R.), LEE (C.C.), GOULD (P.), "GSM, cdmaOne and 3G systems", *John Wiley*, (2001).

[4] STEELE (R.), WILLIAMS (J.E.B.), "Third generation PCN and the intelligent multimode mobile portable', *IEEE Electronics & Communications Eng. J.*, pp. 147-15, (June 1993).

[5] STEELE (R.), "Towards a high capacity digital cellular mobile radio system", *IEE Proc. Pt.5*, pp. 405-415, (Aug. 1985).

[6] STEELE (R.), "Mobile access technologies – what's next", *ACTS Mobile Communications Summit*, Spain, pp. 21-24, (Nov. 27-29, 1996).

[7] HENTSCHEL (T.), HENKER (M.), FETTWEIS (G.), "The digital front-end of software radio terminals", *IEEE Personal Communications*, **6**, n° 4, pp. 40-46, (Aug. 1999).

[8] EL-JABU (B.), STEELE (R.), "Aerial platforms: a promising means of 3G communications", *IEEE Vehicular Technology Conf.*, Houston, (May 1999).

[9] COLELLA (N.J.), MARTIN (J.N.), AKYILDIZ (I.F.), "The HALO Network", *IEEE Communications*, **38**, n° 6, pp. 142-148, (June 2000).

[10] HASE (Y.), MIVRA (R.), OHMORI (S.), "A novel broadband all-wireless access network using stratospheric platforms", *IEEE VTC '98*, pp. 1191-1194.

[11] STEELE (R.), *"A vision of wireless communications in the 21st century"*, Shafi (M.), Mizusawa (J.), Ogose (S.), Hattori (T.), "Wireless Communications in the 21st century', *IEEE PRESS*, (2000).

Index

E

F

G

I

J

L

M